Silicon Photonics

Silicon Photonics
An Introduction

Graham T. Reed
Advanced Technology Institute,
University of Surrey, Guildford, UK

Andrew P. Knights
McMaster University, Hamilton, Ontario, Canada

John Wiley & Sons, Ltd

Other Wiley Editorial Offices

John Wiley & Sons Inc., 111 River Street, Hoboken, NJ 07030, USA

Jossey-Bass, 989 Market Street, San Francisco, CA 94103-1741, USA

Wiley-VCH Verlag GmbH, Boschstr. 12, D-69469 Weinheim, Germany

John Wiley & Sons Australia Ltd, 33 Park Road, Milton, Queensland 4064, Australia

John Wiley & Sons (Asia) Pte Ltd, 2 Clementi Loop #02-01, Jin Xing Distripark, Singapore
129809

John Wiley & Sons Canada Ltd, 22 Worcester Road, Etobicoke, Ontario, Canada M9W 1L1

Wiley also publishes its books in a variety of electronic formats. Some content that appears
in print may not be available in electronic books.

British Library Cataloguing in Publication Data

A catalogue record for this book is available from the British Library

ISBN 10: 0-470-87034-6 (HB)   ISBN 13: 978-0-470-87034-1 (HB)

Typeset in 10.5/13pt Sabon by Laserwords Private Limited, Chennai, India
Printed and bound in Great Britain by TJ International, Padstow, Cornwall
This book is printed on acid-free paper responsibly manufactured from sustainable forestry
in which at least two trees are planted for each one used for paper production.

GTR *dedicates this book to the following people:*

To Alison, Hannah and Matthew for love and inspiration;
To my parents Colleen and John for a lifetime of support;
To Jackie for sibling rivalry!

APK *dedicates this book to Melanie*

Contents

About the Authors

Graham T Reed BSc (Hons), PhD, FIEE, CEng

Silicon Photonics Research Group, Advanced Technology Institute, University of Surrey, UK

Graham Reed is Professor of Optoelectronics at the University of Surrey in the UK. He graduated in 1983 with a First Class Honours degree in Electronic and Electrical Engineering. Subsequently he obtained a PhD in Integrated Optics in 1987. After a brief period as leader of the Electro-Optics Systems Group at ERA Technology Ltd, he joined the University of Surrey in 1989, where he established the Silicon Photonics Group. As such this was one of the pioneering groups in silicon photonics, and has made a significant impact on the state of the art. The group is currently the leading group in the UK in this field, and Professor Reed is acknowledged as the individual who initiated research on silicon photonic circuits and devices in the UK. The aim of the silicon work has been to develop a technology that would have a variety of applications, although telecommunications remains the dominant application area. The work has been carried out with collaborators from all around the world, both from academic and industrial institutions. Professor Reed has published extensively in the international scientific literature, has contributed presentations to numerous international conferences both as a submitting and an invited speaker, and has served on a variety of international committees.

Andrew P Knights BSc (Hons), Ph.D

Department of Engineering Physics, McMaster University, Hamilton, Canada

Andy Knights received his Ph.D. in 1994 in the area of surface and sub-surface material characterisation with beams of low energy positrons and electrons. His subsequent research took him first to the University of Western Ontario where he performed ground-breaking work on the generation and evolution of implant induced defects in silicon, and then to the University of Surrey as part of the EPSRC Ion Beam Centre, researching novel fabrication processes for micro and opto-electronic materials. In 2000 he joined Bookham Technology and worked on a range of silicon-based, highly integrated photonic devices. He played a critical role in the development of the solid-state electronic optical variable attenuator (VOA); the multiplexer + VOA and the monolithically integrated optical detector. In 2003 he moved to McMaster University in Hamilton, Canada, where he holds a faculty position in the Department of Engineering Physics. He currently leads a research group working on the interaction of optical and electrical functionality in silicon-based structures. Dr. Knights has published extensively on semiconductor device design, fabrication and characterization and his work has been presented at international conferences on numerous occasions.

Foreword

'Siliconizing Photonics'

The observation by Gordon Moore in 1965 (now universally referred to as *Moore's law*) that the number of transistors on an integrated circuit would double every couple of years has become a beacon that continues to drive the electronics industry. Integrated circuits have grown exponentially from the 30-transistor devices of 1965 to today's high-end microprocessors exceeding 500 million transistors integrated on a silicon chip the size of your fingernail. Moore's law will continue, with over one billion transistors per chip expected by 2010. Decades of research and manufacturing investment to drive Moore's law has resulted in significant performance gains while simultaneously bringing about significant cost reductions. As an example, in 1968 the cost of a transistor was around one dollar. By 1995, one dollar bought about 3000 transistors. Today, one dollar purchases about five million transistors.

The Internet explosion has changed how we go about our everyday lives. The thirst for information and the need to 'always be connected' is spawning a new era of communications. This new era will continue to spur the need for higher bandwidth technologies to keep pace with processor performance. Because of Moore's law, computing today is limited less by the computer's performance than by the rate at which data can travel between the processor and the outside world. Fiber-optic solutions are replacing copper-based solutions, which can no longer meet the bandwidth and distance requirements needed for worldwide data communications.

Over the last decade, optical communication technologies have migrated steadily from long-haul backbones to the network edge, invading metropolitan area networks (MANs) and campus-level local area networks (LANs). A key inflection point will come with the ability to economically connect central offices to diverse network access points. One of the most important consequences of this migration has been the need to develop more efficient and lower cost optical solutions. The future of optical networking rests on the ability to bring optical technology from the MAN or LAN, into the data center, to the curb, to the home and, if possible, maybe some day directly to the microprocessor.

Today, optical devices are large, bulky and mostly not fabricated from silicon. Most optical components are made from III–V-based compounds such as indium phosphide (InP), gallium arsenide (GaAs) or the electro-optic crystal lithium niobate ($LiNbO_3$). These optical devices are often custom-made and assembled from discrete components. They typically are assembled by hand with very little automation. We are just now beginning to see some standardization occurring in the optical arena. The net result of all this is that these optical devices are relatively expensive. Optical technology to the mass market may happen only if one can bring data-com economics, high-volume manufacturing and assembly to the optical world.

This raises several questions. Can we 'siliconize' photonics? Can we call on the decades of research and manufacturing experience gained from the microelectronics industry and apply it to photonics? Could silicon be used as an alternative to more exotic materials (such as InP or $LiNbO_3$) typically used to produce optical devices? Could one monolithically integrate multiple photonic circuits on a single silicon chip to increase performance while simultaneously reducing cost? Could one implement standardization and high-volume manufacturing techniques to reduce cost? Could we combine electronics with photonics to bring new levels of integration, and possibly a derivative form of Moore's law to photonics?

These are very good questions. Although it is well known that silicon is the optimal material for electronics, only recently has silicon been considered as a practical option for optics. Silicon in fact has many properties conducive to fiber optics. The band gap of silicon ($\sim$1.1 eV) is such that the material is transparent to wavelengths commonly used for optical transport (around 1.3–1.6 μm). One can use standard CMOS-processing techniques to sculpt optical waveguides onto the silicon surface. Similar to an optical fiber, these optical waveguides can be used to confine and direct light as it passes through the silicon.

Due to the wavelengths typically used for optical transport and silicon's high index of refraction, the feature sizes needed for processing these silicon waveguides are on the order of 0.5–1 μm. The lithography requirements needed to process waveguides with these sizes exist today. If we push forward to leading-edge research currently under way in the area of photonic band-gap devices (PBGs), today's state-of-the-art 90-nm fabrication facilities should meet the technical requirements needed for processing PBGs. What this says is that we may already have all or most of the processing technologies needed to produce silicon-based photonic devices for the next decade.

In addition, the same carriers used for the basic functionality of the transistor (i.e. electrons and holes) can be used to modulate the phase of light passing through silicon waveguides and thus produce 'active' rather than passive photonic devices. Finally, if all this remains CMOS-compatible, it could be possible to process transistors alongside photonic devices, the combination of which could bring new levels of performance, functionality, power and size reduction, all at a lower cost.

So the answer is, yes, it is possible to 'siliconize' photonics. Will it happen? The Internet growth engine is alive and well and the optical communications industry will need to move from custom low-volume technologies to high-volume, standardized building blocks. This may happen only if silicon is the material on which we build this photonic technology in the future.

So what of this book? I believe this to be the first comprehensive book on silicon photonics. I found it to be easy to read and believe it will be useful to a wide range of readers. For those studying at a university, at either an undergraduate or graduate level, this book will provide the foundation you need to enter the field. For the corporate researcher working in this new and exciting field, this book will provide an insight into the practical issues and challenges involved with fabricating and producing silicon-based photonic devices.

The authors have put together a comprehensive repository of information, starting from the theoretical fundamentals to outlining the technical and practical issues in producing optical devices in silicon. The book starts with a solid theoretical analysis on the properties of guided waves, optical modes and optical dispersion. The text takes this theory and describes how to translate it into designing waveguides in silicon using a basic rib structure and the parameters required to make these waveguides single-mode. The reader is walked through the processing steps that are needed to produce optical waveguides and photonic devices in silicon, and is exposed to some of the processing tolerances and techniques that effect optical device performance.

A very thorough and detailed analysis for producing an active device (specifically a phase modulator based on current injection) is presented and this gives the reader a good idea of the various parameters that can be varied to improve device performance.

In addition, other devices such as evanescent couplers, AWGs and structures designed for coupling light from small waveguides into an optical fiber are discussed. Finally, there is a technical review of the relatively hotly debated and heavily researched topic ongoing in the world, namely, that of silicon-based light emitters.

This book complements and builds upon over a decade of research that Professor Graham Reed has spent in silicon photonics, and two decades in the wider field of opto-electronics. Professor Reed has been a pioneer in the field of silicon photonics from his initial research on the fundamental properties of waveguiding in silicon to his internationally recognized leading work on devices such as optical modulators, grating couplers and even silicon sensing applications. In addition, Dr Andy Knights brings to the book hands-on fabrication experience in an industrial setting. This experience has given Dr Knights insight into the practical issues associated with the production of silicon-based photonic devices. The combination of experience from Professor Reed and Dr Knights brings together a complementary set of know-how to produce a well-written, comprehensive, introductory book on silicon photonics.

We are at the beginning of a new era in communications, one in which silicon photonics may play a significant role. This book should give readers the foundation on which to participate in this exciting field. When you have read this book I look forward to seeing you some day on this wonderful journey to 'siliconize' photonics.

Dr Mario Paniccia
Dr Paniccia is Director of Optical Technology Development
within the Communications & Interconnect Technology Lab
at Intel Corporation

Acknowledgements

The experience of producing a technical text is a time-consuming business. In particular, the chances of producing an error-free text appear to us to be almost negligible. Therefore we have enlisted help from a number of people to try to minimise the errors in our text, by asking them to read and comment on both the content and accuracy. This has proved invaluable to us, and has identified several omissions and errors. In particular we would like to express our sincere gratitude to the following people:

Dr Mario Paniccia for suggestions and advice on the content of the text, and for his attitude of support and encouragement, which has gone a long way towards getting this book published
Goran Masanovic for meticulous proof reading and diagram production
C. E. Jason Png for additional device simulation, and redrafting some of the diagrams
William Headley, Goran Masanovic, C. E. Jason Png, Seong Phun Chan, Dr Richard Jones, Soon Thor Lim, and Dr Haisheng Rong for finding typographical errors and/or technical errors

All those who allowed us to use their diagrams or other material.

1

Fundamentals

This chapter is intended to reinforce some fundamental concepts before we embark upon a discussion of silicon photonics. When producing such a text, one always faces the philosophical question of 'How much knowledge to assume?' We have chosen to reinforce some fundamental concepts in this chapter, in order to make the text suitable for a wide range of readers. These preliminary discussions will simplify the parameters discussed to some extent to enable the reader to make some rapid progress. However, the more experienced reader should not be put of by the very fundamental nature of this chapter, as it is essential to understand the fundamentals before full use can be made of more advanced material.

1.1 WHAT IS PHASE?

The understanding of phase of an optical wave is fundamental to understanding how waveguides and optical circuits operate. In its simplest terms the concept of phase is very straightforward. If we are dealing with a function of time, the meaning of phase is simply the proportion of a periodic waveform that passes some reference point, after a time t, typically with respect to time $t = 0$.

Let us begin by plotting the function $\sin \theta$, as in Figure 1.1. The phase of the waveform is simply the angle θ. It is implicit that the reference value is 0. Things are a little more complicated if we plot a sinusoidal function that varies with time. For example a plot of $\sin \omega t$ is shown in Figure 1.2, with a time axis. In this case the parameter ω is a constant

Silicon Photonics: An Introduction   Graham T. Reed and Andrew P. Knights
© 2004 John Wiley & Sons, Ltd   ISBN: 0-470-87034-6

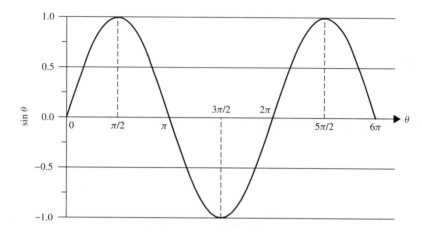

Figure 1.1 Plot of sin θ

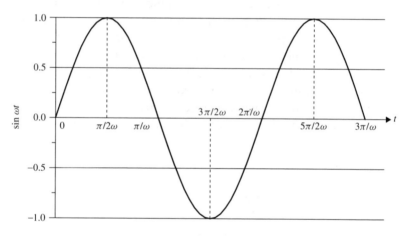

Figure 1.2 Plot of sin ωt

called the *angular frequency*. It is related to the frequency, f, of the waveform by the relationship:

$$\omega = 2\pi f \tag{1.1}$$

Once again the phase of the function is the angle, in this case ωt, but note that because the period of the function is always 2π radians, and the plot is on a time axis and not an ωt axis, the axis crossings now occur every π/ω seconds rather than every π radians.

Let us now consider a propagating optical wave. Such a wave is expressed as:

$$E = E_0 \exp[j(kz \pm \omega t)] \tag{1.2}$$

The exponential function is used for generality, but under some circumstances this can be simplified to a sinusoidal function. For example, let us use a sinusoidal function to help to understand phase a little more:

$$E = E_0 \sin(kz \pm \omega t) \tag{1.3}$$

Following from our previous discussion, the phase is the angle of the sinusoidal function. Thus the phase is:

$$(kz \pm \omega t) \tag{1.4}$$

In this expression, ωt has the same meaning as before, and the term kz is used to describe the progress of the wave in the z direction. k is a constant, similar in concept to ω, which determines the rate of progress of the wave with distance, z.

We can think of the propagating wave, then, as moving in a direction z, but also varying with time. This is very like a wave moving in sea water. The peak of the wave moves through the water at a certain rate (or speed), but at a fixed position in the sea the size of the wave also varies with time.

Therefore equation 1.4 tells us that the wave varies with distance, at a rate determined by the constant k, and varies with time at a rate determined by the constant ω. If we stay at a fixed position and consider the variation of the wave with time, we obtain the variation shown in Figure 1.2.

If we now consider the variation of the wave with distance at a fixed time, we can plot the function sin kz, as in Figure 1.3. Notice that we are now plotting the sinusoidal function against a distance on the z axis. The period of the waveform must still be 2π radians, and therefore the axis crossings now occur every π/k metres rather than every π radians. So k has units of radians per metre, and determines how much of a phase change is experienced per metre of distance that the wave propagates. Therefore k is known as the *propagation constant*. The propagation constant will be discussed in more detail later in this text.

We can also note in passing that the wavelength of the function can be measured directly from Figure 1.3, because the axis has units of metres. The wavelength is the distance between two peaks of the function, which from Figure 1.3 can be seen to be $2\pi/k$. Therefore we can write an expression for wavelength, λ, as:

$$\lambda = \frac{2\pi}{k} \tag{1.5}$$

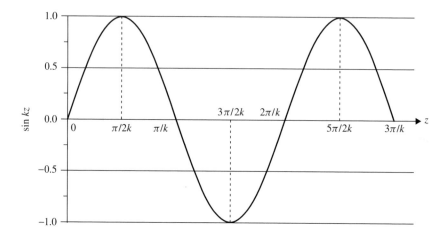

Figure 1.3 Plot of sin kz

Obviously, then, k must be given by:

$$k = \frac{2\pi}{\lambda} \qquad (1.6)$$

If we refer to the phase change over a distance z as ϕ, we can evaluate this phase change as the product of the propagation constant and the distance:

$$\phi = kz \qquad (1.7)$$

In many applications in photonic circuits, we need to measure the phase of one wave with respect to another, perhaps in an interferometer. In such cases the time variation of both waves will be the same, so the time variations are usually neglected. Hence the relative phase difference between the waves is a function of the propagation constant k and the distance of propagation z only. We will use this fact later in the text to explain the operation of several photonic devices.

1.2   WHAT IS POLARISATION?

In order to understand optical waveguides, we must understand what is meant by the polarisation of light. In order to gain such an understanding it is necessary to consider the nature of a lightwave. Light is an electromagnetic wave, which has the characteristic of both electric

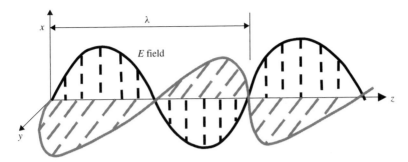

Figure 1.4 Sinusoidal plane wave showing electric and magnetic fields. Reproduced with permission from Palgrave Macmillan

and magnetic fields that vary with time. One of the simplest waves to visualise is a sinusoidal plane wave, shown in Figure 1.4.

The figure shows three characteristics. Firstly, the wave is a *plane* wave because the value of the electric field (shown in the x direction) and the magnetic field (shown in the y direction), are constant throughout any z-plane. Secondly, the wave is *transverse* because both the electric and magnetic fields are transverse to the direction of propagation (z direction). Finally, the wave is *polarised*, because the electric and magnetic field exist in only a single direction. This particular wave is said to be polarised in the x direction because the electric field exists only in the x direction.

Implicit in the previous paragraph is the definition of polarisation. *It is the direction of the electric field associated with the propagating wave.* However, it should always be remembered that there is an associated magnetic field, even though for clarity the magnetic field is not always explicitly considered when discussing propagating waves.

A variety of polarisation configurations are possible in propagating waves, including circular and elliptical polarisation in which the electric vector traces out a circle or an ellipse. However, the most important polarisation configuration for optical circuits is *plane polarised light*, in which all of the waves forming a beam of light have their electric vectors in the same plane, or *unpolarised light* in which the electric vectors are randomly oriented.

We will see later that light in a semiconductor optical waveguide propagates in (almost) plane polarised modes, and the plane in which the light is polarised is either vertical or horizontal with respect to the waveguide surface. It is possible to regard unpolarised light as a combination of these two plane polarised waves, since a wave polarised at an arbitrary angle to the waveguide surface can be resolved into a

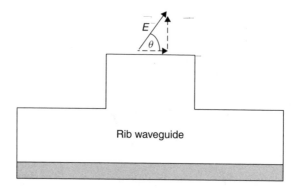

Figure 1.5 Light polarised at an arbitrary angle θ can be resolved into components parallel and perpendicular to the waveguide surface

component parallel to the surface and a component perpendicular to the surface (see Figure 1.5). The issue for optical circuits is that these two components can propagate through the circuit with very different characteristics, resulting in entirely different performance if the light is polarised in one direction rather than the other. This is important because light transmitted to an optical circuit will typically come from an optical fibre, and optical fibres emit light with random polarisation.

1.3 WHAT IS INTERFERENCE?

The phenomenon known as interference is fundamental to the successful operation of many integrated optical devices. Therefore we will review some of the basic characteristics of interference before moving on to discuss specific circuits and devices later in the text.

If two waves are coincident at a given point in space, the fields of the waves add together. Both the electric and magnetic fields of the waves will add together, but it is sufficient to demonstrate the principle of interference by visually summing just one set of fields, say the electric fields. Consider Figure 1.6, for example, in which two sinusoidal electric fields are summed. Note that the sum is carried out several times for fields with a different phase difference between the two waves.

Notice from Figure 1.6 that the fields add in such a way that: (a) if they are in phase constructive interference results; (b) if they are exactly out of phase they cancel one another; and (c) if they are out of phase by an intermediate amount, the resultant field sum is still a sinusoid of the same frequency, but shifted in phase and amplitude.

The interference above has been demonstrated by making two assumptions that have not yet been explicitly stated. Firstly, the two waves that

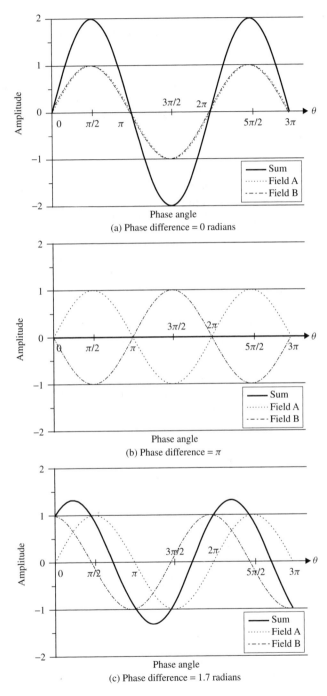

Figure 1.6   Interference of two waves of varying phase difference

are adding are either the electric fields or the magnetic fields. The fact that they are adding implies that they are *aligned in terms of polarisation*. If two fields are not of the same polarisation, the electric and magnetic vectors are not aligned, and the fields will not interfere. If the waves are both linearly polarised, but the polarisation directions are different, only the components of each field that have common polarisation will interfere. In the particular case when the two waves have polarisation directions that are 90° to one another (orthogonal), no interference will occur at all.

Secondly, the two waves will only interfere if they are *coherent*. The term 'coherence' describes the phase relationship between two or more waves. If two waves are coherent they have a constant phase relationship, which allows the simple addition of fields that we have just observed. This will often mean in practice that the two waves are derived from the same optical source. If on the other hand the two waves have no fixed phase relationship, they are said to be *incoherent*, and we cannot simply add the fields because the phase relationship will be random, and will vary with time. In the case of incoherent waves, the fields do not add, but the total power in two incoherent waves is additive. This contrasts to the case of the coherent waves whose total power is proportional to the square of the total electric field (this is because the power in any wave is proportional to the square of the electric field).

This description of coherence is incomplete without considering the intermediate condition of 'partial coherence'. This is important in practice, because no light source is truly monochromatic, and hence the phase relationship between waves, even from the same source, will eventually break down after propagating a certain distance. This is because the finite spectral width of the light source means that the component wavelengths within the spectral width all develop slightly different phase relationships with propagation distance. Therefore the *coherence length* is a useful parameter. The coherence length is the distance over which the light retains coherence. Alternatively this can be regarded as the maximum path difference two waves from the same source can experience such that interference can still occur. The mathematical definition of coherence length, L_c, is:

$$L_c = \frac{c}{\Delta f} \qquad (1.8)$$

where c is the velocity of light, and Δf is the spectrum of frequencies contained within the light source.

Of course, the quality of interference does not change from very good to nonexistent as the coherence length is exceeded, but gradually reduces with increasing propagation distance. This manifests itself in practice as the 'contrast' between constructive and destructive interference getting worse with propagation distance. Consequently, when using an interferometer, it is important to ensure that the path length difference between the interference waves is much less than the coherence length of the optical source.

It is instructive to consider the coherence length of different types of optical sources, to gain a 'feel' for the sort of source quality that is required for a given application. A gas laser found on the bench of an optical laboratory, such as a helium–neon laser, may have a coherence length measured in metres, tens of metres, or even hundreds of metres, whereas an LED (light-emitting device) with a spectral width of, say, 100 nm has a coherence length approximately 6 μm. Since the coherence length represents the degree of path difference allowable in an interferometer, clearly it would be impossible to fabricate an interferometer in free space using the latter device.

2

The Basics of Guided Waves

2.1 THE RAY OPTICS APPROACH TO DESCRIBING PLANAR WAVEGUIDES

The study of light is a study of electromagnetic waves. Consequently the photonics engineer will inevitably encounter electromagnetic theory during his or her career. However, the rigours of Maxwell's equations are not always required for all applications, and one can make significant progress in understanding the basics of guided wave propagation with more simple models. Therefore this section introduces the well-known ray optical model, and uses this model to investigate a number of important phenomena in simple optical waveguides. Later in the text we will build on these concepts to describe more complex structures.

Firstly consider a light ray (E_i) propagating in a medium with refractive index n_1, and impinging on the interface between two media, at an angle θ_1, as shown in Figure 2.1. At the interface the light is partially transmitted (E_t) and partially reflected (E_r). The relationship between the refractive indices n_1 and n_2, and the angles of incidence (θ_1) and refraction (θ_2), is given by *Snells law*:

$$n_1 \sin \theta_1 = n_2 \sin \theta_2 \qquad (2.1)$$

In Figure 2.1, the refractive index of medium 1 (n_1) is clearly higher than the refractive index of medium 2 (n_2), as the angle θ_2 is greater than θ_1. Consequently as θ_1 is increased, θ_2 will approach 90°. For some angle θ_1, the corresponding angle θ_2 will reach 90°, and hence Snell's law simplifies to:

$$n_1 \sin \theta_1 = n_2 \qquad (2.2)$$

Silicon Photonics: An Introduction   Graham T. Reed and Andrew P. Knights
© 2004 John Wiley & Sons, Ltd   ISBN: 0-470-87034-6

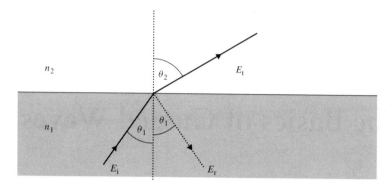

Figure 2.1   Light rays refracted and reflected at the interface of two media

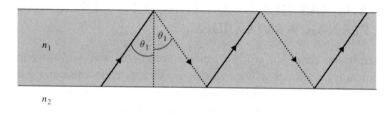

Figure 2.2   Total internal reflection at two interfaces, demonstrating the concept of a waveguide

Hence we can define this critical angle θ_c as:

$$\sin \theta_c = \frac{n_2}{n_1} \qquad (2.3)$$

For angles of incidence greater than this critical angle, no light is transmitted and total internal reflection (TIR) occurs. If we now consider a second interface below the first, at which the wave also experiences total internal reflection, we can understand the concept of a waveguide, as the light is confined to the region of refractive index n_1, and propagates to the right (Figure 2.2). This simplistic approach suggests, however, that the waveguide will support propagation at any angle greater than the critical angle. In the next section we shall see that this is not the case.

2.2   REFLECTION COEFFICIENTS

To fully understand the behaviour of the optical waveguide it is necessary to use electromagnetic theory. However, by enhancing our ray model a little we can improve our understanding significantly.

Thus far we have implicitly considered reflection/transmission at an interface to be 'partial' reflection and 'partial' transmission, without defining the term 'partial'. With reference to Figure 2.1, consider the reflection and transmission of the wave at a single interface. It is well known that the reflected and transmitted waves can be described by the *Fresnel formulae*. For example, the reflected wave will have complex amplitude E_r at the interface, related to the complex amplitude E_i of the incident wave by:

$$E_r = rE_i \qquad (2.4)$$

where r is a complex reflection coefficient. The reflection coefficient is a function of both the angle of incidence and the polarisation of light. Hence before proceeding further we must define the polarisation with respect to the interface.

The electric and magnetic fields of an electromagnetic wave are always orthogonal to one another, and both are orthogonal to the direction of propagation. Hence propagating electromagnetic waves are also referred to as *transverse electromagnetic waves*, or TEM waves. The polarisation of a wave is deemed to be the direction of the electric field associated with the wave. For the purposes of this discussion we wish to consider cases where either the electric field, or the magnetic field, is perpendicular to the plane of incidence (i.e. the plane containing the wave normal and the normal to the interface).

The *transverse electric* (TE) condition is defined as the condition when the electric fields of the waves are perpendicular to the plane of incidence. This is depicted in Figure 2.3. Correspondingly, the *transverse magnetic* (TM) condition occurs when the magnetic fields are perpendicular to the plane of incidence. As previously mentioned, the Fresnel formulae describe the reflection coefficients r_{TE} and r_{TM}, usually written as follows.

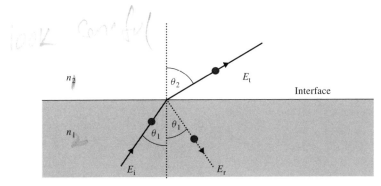

Circles ● indicate that the electric fields are vertical (i.e. coming out of the plane of the paper)

Figure 2.3 Orientation of electric fields for TE incidence at the interface between two media

For TE polarisation:

$$r_{TE} = \frac{n_1 \cos\theta_1 - n_2 \cos\theta_2}{n_1 \cos\theta_1 + n_2 \cos\theta_2} \tag{2.5a}$$

Similarly for TM polarisation:

$$r_{TM} = \frac{n_2 \cos\theta_1 - n_1 \cos\theta_2}{n_2 \cos\theta_1 + n_1 \cos\theta_2} \tag{2.6a}$$

Using Snell's law (equation 2.1), equations 2.5a and 2.6a can be re-written as:

$$r_{TE} = \frac{n_1 \cos\theta_1 - \sqrt{n_2^2 - n_1^2 \sin^2\theta_1}}{n_1 \cos\theta_1 + \sqrt{n_2^2 - n_1^2 \sin^2\theta_1}} \tag{2.5b}$$

$$r_{TM} = \frac{n_2^2 \cos\theta_1 - n_1\sqrt{n_2^2 - n_1^2 \sin^2\theta_1}}{n_2^2 \cos\theta_1 + n_1\sqrt{n_2^2 - n_1^2 \sin^2\theta_1}} \tag{2.6b}$$

When the angle of incidence is less than the critical angle, only partial reflection occurs and the reflection coefficient is real. However, when the critical angle is exceeded we have total internal reflection. We can see from equations 2.5b and 2.6b that the term inside the square-root becomes negative. Clearly this implies that:

$$|r| = 1 \tag{2.7}$$

and that r is also complex and hence a phase shift is imposed on the reflected wave.

Hence we may denote this by:

$$r = \exp(j\phi) \tag{2.8}$$

where ϕ_{TE} and ϕ_{TM} are given by:

$$\phi_{TE} = 2\tan^{-1}\frac{\sqrt{\sin^2\theta_1 - \left(\frac{n_2}{n_1}\right)^2}}{\cos\theta_1} \tag{2.9}$$

$$\phi_{TM} = 2\tan^{-1}\frac{\sqrt{\frac{n_1^2}{n_2^2}\sin^2\theta_1 - 1}}{\frac{n_2}{n_1}\cos\theta_1} \tag{2.10}$$

Note, however, that we have now defined reflection coefficients without too much discussion of the incident and reflecting waves. In fact it is important to note that the reflection coefficients, as defined, relate the relative amounts of the fields (not the powers) which are reflected. In practical terms the field is difficult to measure directly, so we are more interested in the power (or intensity) that is reflected or transmitted. In an electromagnetic wave, the propagation of power is described by the *Pointing vector*, usually denoted as S (measured in W/m^2). The Pointing vector is defined as the vector product of the electric and magnetic vectors, and hence indicates both the magnitude and direction of power flow. One of the most common ways of expressing the Pointing vector is:

$$S = \frac{1}{Z}E^2 = \sqrt{\frac{\varepsilon_m}{\mu_m}}E^2 \qquad (2.11)$$

where E is electric field, ε_m is the permittivity of the medium, μ_m is the permeability of the medium, and Z is the impedance of the medium. We can then define a reflectance, R, which relates the incident and reflected powers associated with the waves as:

$$R = \frac{S_r}{S_i} = \frac{E_r^2}{E_i^2} = r^2 \qquad (2.12)$$

2.3 PHASE OF A PROPAGATING WAVE AND ITS WAVEVECTOR

If we enhance our simple ray model a little further we can understand more about the propagation of light in simple waveguides. We will also be able to improve on the simple approach to the propagation constant in Chapter 1. Thus far the ray has simply represented the *direction* of propagating light. We now need briefly to consider a propagating electromagnetic wave. In common with electrical circuit theory, it is convenient to use the exponential form of the propagating wave, to make a sufficiently general point. In common with our approach in Chapter 1, let the electric and magnetic fields associated with a propagating wave be described respectively as:

$$E = E_0 \exp[j(kz \pm \omega t)] \qquad (2.13)$$

$$H = H_0 \exp[j(kz \pm \omega t)] \qquad (2.14)$$

where z is merely the direction of propagation (for the sake of argument). This means that the phase of the wave (as in Chapter 1) is:

$$\phi = kz \pm \omega t \qquad (2.15)$$

It is clear that the phase varies with time (t), and with distance (z). These variations are quantified by taking the time derivative and the spatial derivative:

$$\left| \frac{\partial \phi}{\partial t} \right| = \omega = 2\pi f \qquad (2.16)$$

where ω is angular frequency (measured in rad/s), and f is frequency (Hz). Both angular frequency and frequency in Hertz are familiar to engineers, and describe how the phase of the wave varies with time. A similar expression can be found to relate phase to propagation distance by taking the spatial derivative of equation 2.15:

$$\frac{\partial \phi}{\partial z} = k \qquad (2.17)$$

where k is the wavevector or the propagation constant in the direction of the wavefront. It is related to wavelength, λ, by:

$$k = \frac{2\pi}{\lambda} \qquad (2.18)$$

In free space, k is usually designated k_0. Hence k and k_0 are related by n, the refractive index of the medium:

$$k = nk_0 \qquad (2.19)$$

Hence in free space:

$$k_0 = \frac{2\pi}{\lambda_0} \qquad (2.20)$$

2.4   MODES OF A PLANAR WAVEGUIDE

Having defined the wavevector we can now further consider propagation in a waveguide. The simplest optical waveguide is the planar type, depicted in Figure 2.2. This diagram is reproduced in Figure 2.4a, with the addition of axes x, y and z.

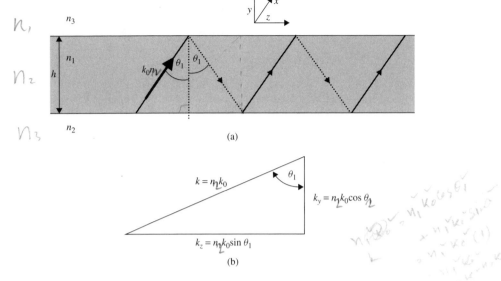

Figure 2.4 (a) Propagation in a planar waveguide. (b) The relationship between propagation constants in the y, z and wavenormal directions. Adapted with permission from Artech House Publishing, Norwood, MA, USA, www.artechhouse.com

Let the waveguide height be h, and propagation be in the z direction, with the light being confined in the y direction by total internal reflection. The zig-zag path indicated in the figure now means more than it did previously. It represents the direction of the wavenormals as the waves propagate through the waveguide, with wavevector $k\ (= k_0 n_1)$. We can use an associated diagram (Figure 2.4b) to explain this further, in which we decompose the wavevector k into two components, in the y and z directions. By simple trigonometry,

$$k_z = n_1 k_0 \sin \theta_1 \quad (2.21)$$
$$k_y = n_1 k_0 \cos \theta_1 \quad (2.22)$$

Having determined the propagation constant in the y direction, we can imagine a wave propagating in the y direction. Since this theoretical wave will be reflected at each interface, there will potentially be a *standing wave* across the waveguide in the y direction. Hence we can sum all the phase shifts introduced in making one complete 'round trip' across the waveguide and back again. For a waveguide thickness h (and hence a traversed distance of $2h$), using equation 2.17, a phase shift is introduced of:

$$\phi_h = 2k_y h = 2k_0 n_1 h \cos \theta_1 \quad (2.23)$$

From equations 2.8–2.10, it is clear that phase changes are introduced upon reflection at upper and lower waveguide boundaries. Let us refer to these phase shifts as ϕ_u and ϕ_ℓ respectively. Hence the total phase shift is:

$$*\phi_t = 2k_0 n_1 h \cos\theta_1 - \phi_u - \phi_\ell \tag{2.24}$$

For consistency (i.e. the preservation of a wave across the waveguide), this total phase shift must be a multiple of 2π; hence:

$$2k_0 n_1 h \cos\theta_1 - \phi_u - \phi_\ell = 2m\pi \tag{2.25a}$$

where m is an integer. This equation can be reduced to an equation in θ, by substituting for ϕ_u and ϕ_ℓ from equations 2.9 and 2.10.

Before we do that, however, notice that because m is an integer, there will be a series of discrete angles θ for which equation 2.25a can be solved, corresponding to integral values of m. For each solution there will therefore be a corresponding propagation constant in the y and z directions (for each polarisation). This shows us that light cannot propagate at any angle θ, but only at one of the allowed discrete angles. Each allowed solution is referred to as a *mode of propagation*, and the *mode number* is given by the value of integer m. The modes are identified using a notation utilising the polarisation and the mode number. For example, the first TE mode, or *fundamental mode*, will be described as TE_0. Higher-order modes are correspondingly described using the appropriate value of m. There is also a further limit on m, indicating that there is a limit to the number of modes that can propagate in a given waveguide structure. The limiting conditions correspond to the propagation angle, θ_1, becoming less than the critical angle at either the upper or lower waveguide interface. We will consider this further in the next few sections, as we discuss the modes in the waveguides more comprehensively.

2.4.1 The Symmetrical Planar Waveguide

Having established equation 2.25a, which describes the discrete nature of waveguide modes, let us now discuss solving the equation. The planar waveguide we have discussed thus far was described by Figure 2.4, having an upper cladding of refractive index n_3 and a lower cladding refractive index of n_2. In the case where $n_2 = n_3$, the waveguide is

---

* by convention ϕ_u and ϕ_ℓ are shown as negative in this equation, whereas in reality the negative signs originate from equations 2.9 and 2.10.

said to be symmetrical since the same boundary conditions will apply at the upper and lower interfaces. In terms of solving equation 2.25a, this means that $\phi_u = \phi_\ell$. Therefore using equation 2.9 (assuming TE polarisation), equation 2.25a becomes:

$$2k_0 n_1 h \cos\theta_1 - 4\tan^{-1}\left[\frac{\sqrt{\sin^2\theta_1 - (n_2/n_1)^2}}{\cos\theta_1}\right] = 2m\pi \qquad (2.25b)$$

This can be rearranged as:

$$\tan\left[\frac{k_0 n_1 h \cos\theta_1 - m\pi}{2}\right] = \left[\frac{\sqrt{\sin^2\theta_1 - (n_2/n_1)^2}}{\cos\theta_1}\right] \qquad (2.26)$$

The only variable in equation 2.26 is θ_1, so solving the equation yields the propagation angle. It is also a straightforward calculation to find the propagation constants associated with the mode in question. Using equation 2.10 for the phase shift, the corresponding TM equation is:

$$\tan\left[\frac{k_0 n_1 h \cos\theta_1 - m\pi}{2}\right] = \left[\frac{\sqrt{(n_1/n_2)^2 \sin^2\theta_1 - 1}}{(n_2/n_1)\cos\theta_1}\right] \qquad (2.27)$$

It is often convenient to have an approximate idea of the number of modes supported by a waveguide. Consider equation 2.26, the eigenvalue equation for TE modes. We know that the minimum value that θ_1 can take corresponds to the critical angle θ_c. From equation 2.3:

$$\sin\theta_c = \frac{n_2}{n_1} \qquad (2.28)$$

Hence the right-hand side of equation 2.26 reduces to zero. Since θ_1 decreases with mode number, the minimum value of θ_1 ($= \theta_c$) corresponds to the highest possible order mode. At $\theta_1 = \theta_c$, equation 2.26 reduces to:

$$\frac{k_0 n_1 h \cos\theta_c - m_{\max}\pi}{2} = 0 \qquad (2.29a)$$

Rearranging for m, the mode number is:

$$m_{max} = \frac{k_0 n_1 h \cos \theta_c}{\pi} \tag{2.29b}$$

We can see that if we evaluate m_{max}, and find the nearest integer that is less than the evaluated m_{max}, then this will be the highest-order mode number, which we can denote as $[m_{max}]_{int}$. The number of modes will actually be $[m_{max}]_{int} + 1$, since the lowest-order mode (usually called the fundamental mode) has a mode number $m = 0$.

In passing we can also note a very interesting characteristic of the symmetrical waveguide, from equations 2.26 and 2.27, the eigenvalue equations for TE and TM polarisations. Both equations allow a solution when $m = 0$. This is because the term in the square-root on the right-hand side of the equation is always positive, and hence the square-root is always a real number. This in turn is because θ_1 is always greater than the critical angle, and hence $\sin^2 \theta_1$ is always greater than $\sin^2 \theta_c$ ($= n_2^2/n_1^2$). This implies that the lowest-order mode is always allowed. This means that the fundamental mode will always propagate, and the waveguide is never 'cut-off'. We will see shortly that this is not the case for the asymmetrical waveguide.

2.4.2   The Asymmetrical Planar Waveguide

Let us now consider the slightly more complex asymmetrical planar waveguide. This is shown in Figure 2.5. In this case $n_2 \neq n_3$.

We can approach a simple analysis in the same way as we did for the symmetrical planar waveguide, but the phase change on reflection at the upper and lower waveguide boundaries will not be the same. This time

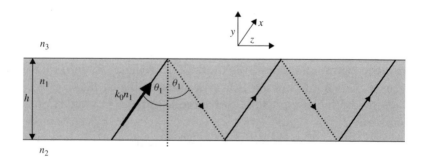

Figure 2.5   Propagation in an asymmetrical planar waveguide

the eigenvalue equation for TE modes becomes:

$$[k_0 n_1 h \cos\theta_1 - m\pi] = \tan^{-1}\left[\frac{\sqrt{\sin^2\theta_1 - (n_2/n_1)^2}}{\cos\theta_1}\right]$$

$$+ \tan^{-1}\left[\frac{\sqrt{\sin^2\theta_1 - (n_3/n_1)^2}}{\cos\theta_1}\right] \qquad (2.30)$$

Once again this equation can be solved (numerically or graphically) to find propagation angle θ_1, for a given value of m. Note, however, that there is not always a solution to equation 2.30 for $m = 0$, because it is possible for one of the terms within the square-roots on the right-hand side to be negative. This is because the critical angle for the waveguide as a whole will be determined by the larger of the critical angles of the two waveguide boundaries. Obviously for total internal reflection at both waveguide boundaries, the propagating mode angle must be greater than both critical angles. However, the mode angle of the waveguide may not satisfy the condition for both critical angles to be exceeded if the guide is too thin (h is small), or if the refractive index difference between core and claddings is too small. Let us consider this in more detail by solving the eigenvalue equation for TE modes, equation 2.30.

2.4.3   Solving the Eigenvalue Equations for Symmetrical and Asymmetrical Waveguides

The eigenvalue equation can be solved in a number of ways, typically numerically. However, it can be more instructive to solve the equation graphically. Firstly we need to choose some waveguide parameters, so let $n_1 = 1.5, n_2 = 1.49, n_3 = 1.40, \lambda_0 = 1.3\,\mu\text{m}$, and $h = 0.3\,\mu\text{m}.^*$ Whilst these refractive indices are not representative of a silicon waveguide, they will enable us to make a point about modal characteristics.

Figure 2.6 shows a graphical solution of the TE eigenvalue equation (equation 2.30) for $m = 0$. The three terms in the equation are plotted separately for clarity. The two phase change terms, ϕ_u and ϕ_ℓ, are added and also shown, representing the right-hand side of equation 2.30. Where this curve intersects with the left-hand side of the equation is the solution for the mode angle, θ_1. This is shown on the diagram. Notice that this solution occurs at an angle θ_1 that lies between θ_u and θ_ℓ,

---

* Note that out simple model will become less accurate for dimensions smaller than the wavelength, so care must be taken when using the results.

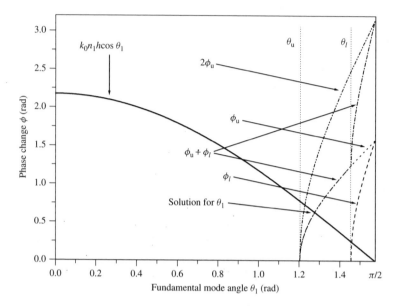

Figure 2.6   Solution of the eigenvalue equation for $m = 0$

the critical angles for the upper and lower interfaces respectively (also shown on the diagram by the dotted vertical lines). Thus the conditions for total internal reflection are not met, and the waveguide is cut-off. If we increase the waveguide thickness, h, the curves would eventually intersect at an angle greater than both θ_u and θ_ℓ. This confirms that the asymmetrical waveguide is cut off for some conditions. Alternatively if we consider the symmetrical guide (with $n_2 = n_3 = 1.4$), the right-hand side of the equation becomes $2\phi_u$ (also plotted). It is clear that the solution will always be at an angle greater than θ_u, confirming that the symmetrical waveguide is never cut off.

The corresponding graphical solution for a silicon waveguide is shown in Figure 2.7. The waveguide parameters are $n_1 = 3.5$ (silicon), $n_2 = 1.5$ (silicon dioxide), $n_3 = 1.0$ (air), $\lambda_0 = 1.3\,\mu m$, and $h = 0.15\,\mu m$. It is interesting to note there is relatively little difference between the symmetrical and the asymmetrical silicon waveguide, because typical cladding material for silicon is silicon dioxide (SiO_2), which has a refractive index of 1.5. Hence either cladding with air or SiO_2 means that the cladding is very different from the core refractive index of approximately 3.5.

2.4.4   Monomode Conditions

It is often convenient for a waveguide to support only a single mode, for a given polarisation of light. Such a waveguide is referred to as *monomode*.

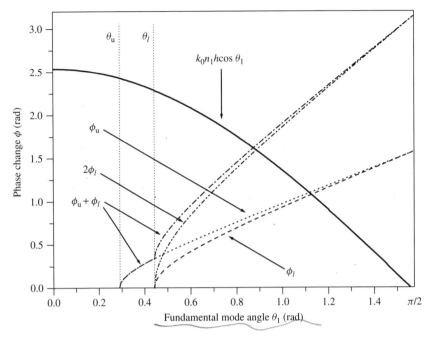

Figure 2.7 Solution of the eigenvalue equation for $m = 0$ (silicon-on-insulator)

We will return to this point later in this text. For now, consider again the TE polarisation eigenvalue equation 2.26 for a symmetrical waveguide, reproduced below:

$$\tan\left[\frac{k_0 n_1 h \cos \theta_1 - m\pi}{2}\right] = \left[\frac{\sqrt{\sin^2 \theta_1 - (n_2/n_1)^2}}{\cos \theta_1}\right] \qquad (2.31)$$

Considering the second mode, with mode number $m = 1$, the limiting condition for this mode is that the propagation angle is equal to the critical angle. Under these circumstances the square-root term reduces to zero. Since the propagation angle for the second mode will be less than for the fundamental mode ($m = 0$), for all angles greater than this critical angle the waveguide will be monomode. Thus equation 2.31 reduces to:

$$\tan\left[\frac{k_0 n_1 h \cos \theta_c - \pi}{2}\right] = 0 \qquad (2.32a)$$

so that

$$\cos \theta_c = \frac{\pi}{k_0 n_1 h} = \frac{\lambda_0}{2 n_1 h} \qquad (2.32b)$$

Hence for monomode conditions:

$$\theta_c < \cos^{-1}\left(\frac{\lambda_0}{2n_1 h}\right) \tag{2.33}$$

Similar expressions can be derived from the other eigenvalue equations.

2.4.5   Effective Index of a Mode

Thus far we have used a simple model to look at the characteristics of a planar waveguide in order to gain an understanding of modes. We can take the model still further to understand a little more. Earlier we briefly discussed the idea of propagation constants in the y and z directions. These were defined in equations 2.21 and 2.22 as:

$$k_z = n_1 k_0 \sin\theta_1 \tag{2.34}$$

$$k_y = n_1 k_0 \cos\theta_1 \tag{2.35}$$

The propagation constant in the z direction is of particular interest as it indicates the rate at which the wave propagates in the z direction. k_z is often replaced in many texts by the variable β, and the two terms are used interchangeably.

We can now define a parameter N, called the *effective index of the mode*, such that:

$$N = n_1 \sin\theta_1 \tag{2.36}$$

Then equation 2.34 becomes:

$$k_z = \beta = Nk_0 \tag{2.37}$$

Note the similarity with equation 2.19. This is equivalent to thinking of the mode as propagating straight down the waveguide, without 'zig-zagging' back and forth, with refractive index N. Having defined β, we can immediately consider the range of values that β can take.

The lower bound on β is determined by the critical angles of the waveguide. For the asymmetrical waveguide the smaller of the two critical angles is usually defined by the upper cladding layer, as this usually has a lower refractive index than the lower cladding layer. In the case of silicon, the lowest possible value of the refractive index of the upper cladding is 1.0, if the upper cladding is air. This means that total internal reflection is limited by the lower cladding, which will have a larger critical angle that needs to be exceeded for total internal reflection to occur. Hence $\theta_1 \geq \theta_\ell$. Hence the lower bound on β is given by:

$$\beta \geq n_1 k_0 \sin \theta_\ell = k_0 n_2 \tag{2.38}$$

The upper bound on β is governed by the maximum value of θ, which is clearly 90°. In this case $\beta = k = n_1 k_0$. Hence:

$$k_0 n_1 \geq \beta \geq k_0 n_2 \tag{2.39}$$

Using equation 2.37 we find that:

$$n_1 \geq N \geq n_2 \tag{2.40}$$

2.5 A TASTE OF ELECTROMAGNETIC THEORY

Thus far we have essentially avoided electromagnetic theory. This section is included to gain a flavour of the electromagnetic model, and perhaps more importantly, relate it to the ray model. It is a standard approach, available in many textbooks, but helps our simplified approach at this point in the text. It is worth noting, however, that so far we have occasionally introduced aspects of the electromagnetic model without direct reference, or indeed much justification. For example in section 2.3, we merely assumed electric and magnetic field distributions in order to understand propagation constants. Maxwell's equations relate the electric field E (V/m), magnetic field H (A/m), charge density ρ (C/m^3), and current density J (A/cm^2). They are commonly written in the following form:

$$\nabla.\mathbf{D} = \rho \tag{2.41}$$

$$\nabla.\mathbf{B} = 0 \tag{2.42}$$

$$\nabla \times \mathbf{E} = -\partial \mathbf{B}/\partial t \tag{2.43}$$

$$\nabla \times \mathbf{H} = \mathbf{J} + \frac{\partial \mathbf{D}}{\partial t} \tag{2.44}$$

where the del operator, ∇, is given by:

$$\nabla = \left(\frac{\partial \mathbf{i}}{\partial x}, \frac{\partial \mathbf{j}}{\partial y}, \frac{\partial \mathbf{k}}{\partial z} \right) \tag{2.45}$$

where $\mathbf{i}$, $\mathbf{j}$ and $\mathbf{k}$ are unit vectors in the x, y and z directions respectively. Note in particular that $\mathbf{i}$ should not be confused with current, nor $\mathbf{j}$ with $\sqrt{(-1)}$, nor $\mathbf{k}$ with a propagation constant.

The electric field $\mathbf{E}$ and magnetic field $\mathbf{H}$ are related to the electric flux density $\mathbf{D}$ and magnetic flux density $\mathbf{B}$ (assuming a lossless medium) by:

$$\mathbf{D} = \varepsilon_{\mathrm{m}}\mathbf{E} \tag{2.46}$$

$$\mathbf{B} = \mu_{\mathrm{m}}\mathbf{H} \tag{2.47}$$

where ε_{m} is the permittivity of the medium and μ_{m} is the permeability of the medium. In reality, the situation is a little more complicated for silicon, but the principles of the approach we will follow are identical. We will also need *Ohm's law*, expressed in terms of current density:

$$\mathbf{J} = \sigma\mathbf{E} \tag{2.48}$$

where σ is the conductivity (measured in $1/\Omega\mathrm{m}$).

Firstly we will use the del operator on equation 2.43 (take the curl, or vector cross-product):

$$\nabla \times \nabla \times \mathbf{E} = -\frac{\partial}{\partial t}(\nabla \times \mathbf{B}) \tag{2.49}$$

Now substitute for $\mathbf{B}$ from equation 2.47:

$$\nabla \times \nabla \times \mathbf{E} = -\mu_{\mathrm{m}}\frac{\partial}{\partial t}(\nabla \times \mathbf{H}) \tag{2.50}$$

Now substitute for $\nabla \times \mathbf{H}$ from equation 2.44:

$$\nabla \times \nabla \times \mathbf{E} = -\mu_{\mathrm{m}}\left[\frac{\partial \mathbf{J}}{\partial t} + \frac{\partial^2 \mathbf{D}}{\partial t^2}\right] \tag{2.51}$$

Now substituting for $\mathbf{J}$ (from equation 2.48), and for $\mathbf{D}$ (from equation 2.46), equation 2.51 becomes:

$$\nabla \times \nabla \times \mathbf{E} = -\mu_{\mathrm{m}}\left[\sigma\frac{\partial \mathbf{E}}{\partial t} + \varepsilon_{\mathrm{m}}\frac{\partial^2 \mathbf{E}}{\partial t^2}\right] \tag{2.52}$$

If we now apply the vector identity:

$$\nabla \times \nabla \times \mathbf{E} = \nabla(\nabla.\mathbf{E}) - \nabla^2\mathbf{E} \tag{2.53}$$

to the left-hand side of equation 2.52, and make the simplifying assumption that $\rho = 0$, then $\nabla.\mathbf{E} = 0$. If we further assume that $\sigma = 0$, we have:

$$\nabla^2 \mathbf{E} = \mu_m \varepsilon_m \frac{\partial^2 \mathbf{E}}{\partial t^2} \qquad (2.54)$$

where ∇^2 is the Laplacian operator, given by:

$$\nabla^2 = \left(\frac{\partial^2}{\partial x^2} + \frac{\partial^2}{\partial y^2} + \frac{\partial^2}{\partial z^2} \right) \qquad (2.55)$$

Equation 2.54 is known as the *wave equation*, and describes the propagation of electromagnetic waves. Solving the wave equation for waveguides describes the modes of the waveguides mathematically. This has the explicit advantage over the ray optical model of describing the fields of the modes, allowing us the visualise the modes, and vitally, to see how one field relates to another, for example in an optical coupling process from one waveguide to another. Note that for the purposes of this text we have made the simplification that the waveguide material is non-conducting (i.e. $\sigma = 0$), in order to simplify the mathematics. This will enable us to produce straightforward approximate solutions to demonstrate the usefulness of the approach.

Since the velocity of a wave is given by:

$$v^2 = \frac{1}{\mu_m \varepsilon_m} \qquad (2.56)$$

then the wave equation is sometimes written as:

$$\nabla^2 \mathbf{E} = \frac{1}{v^2} \frac{\partial^2 \mathbf{E}}{\partial t^2} \qquad (2.57)$$

Clearly, similar equations can be derived for **H**.

2.6 SIMPLIFYING AND SOLVING THE WAVE EQUATION

In this section we will use the simple waveguide structure of the planar waveguide to simplify the wave equation discussed in section 2.5. Using one polarisation corresponding to TE modes, we will then solve the wave equation. The process is identical for TM modes, resulting in a similar solution as expected. Since this text is intended for readers relatively new to the subject, this section of the text is explained in detail and may be a little too pedestrian for the more experienced.

Let us begin by expressing the electric field as its constituent parts in cartesian coordinates:

$$\mathbf{E} = E_x \mathbf{i} + E_y \mathbf{j} + E_z \mathbf{k} \tag{2.58}$$

Now let us rewrite the wave equation, equation 2.54, relabelled as 2.59:

$$\nabla^2 \mathbf{E} = \mu_m \varepsilon_m \frac{\partial^2 \mathbf{E}}{\partial t^2} \tag{2.59}$$

If we expand the left-hand side we have:

$$
\begin{aligned}
\nabla^2 \mathbf{E} &= \frac{\partial^2 \mathbf{E}}{\partial x^2} + \frac{\partial^2 \mathbf{E}}{\partial y^2} + \frac{\partial^2 \mathbf{E}}{\partial z^2} \\
&= \left(\frac{\partial^2 E_x}{\partial x^2} + \frac{\partial^2 E_x}{\partial y^2} + \frac{\partial^2 E_x}{\partial z^2} \right) \mathbf{i} + \left(\frac{\partial^2 E_y}{\partial x^2} + \frac{\partial^2 E_y}{\partial y^2} + \frac{\partial^2 E_y}{\partial z^2} \right) \mathbf{j} \\
&+ \left(\frac{\partial^2 E_z}{\partial x^2} + \frac{\partial^2 E_z}{\partial y^2} + \frac{\partial^2 E_z}{\partial z^2} \right) \mathbf{k}
\end{aligned}
\tag{2.60}
$$

Since, in TEM waves, the electric and magnetic fields are orthogonal, in many situations appropriate selection of axes can simplify the mathematics. We have already made assumptions in the description above that the fields can be described as having components in the x, y and/or z directions. Recalling our previous ray model diagram of the planar waveguide (Figure 2.5), we can now consider the polarisation of the fields associated with the wave equation. With reference to Figure 2.8, recall that TE polarisation means that the electric field exists only in the x direction. Furthermore, the field is uniform in the x direction, since the

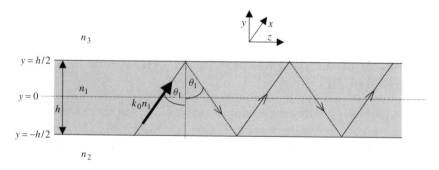

Figure 2.8 Propagation in an asymmetrical planar waveguide

planar waveguide is assumed to be infinite in that direction. We will also limit propagation to the z direction. Hence equation 2.60 reduces to:

$$\nabla^2 \mathbf{E} = \left(\frac{\partial^2 E_x}{\partial y^2} + \frac{\partial^2 E_x}{\partial z^2}\right)\mathbf{i} \qquad (2.61)$$

This means that the vector wave equation 2.59 reduces to a scalar wave equation (since polarisation in the x direction is assumed). This is, substituting 2.61 into 2.59 gives:

$$\frac{\partial^2 E_x}{\partial y^2} + \frac{\partial^2 E_x}{\partial z^2} = \mu_m \varepsilon_m \frac{\partial^2 E_x}{\partial t^2} \qquad (2.62)$$

Since there is only electric field in the x direction, we can write the equation of this field as:

$$E_x = E_x(y)e^{-j\beta z}e^{j\omega t} \qquad (2.63)$$

This means that there is a field directed (polarised) in the x direction, with a variation in the y direction yet to be determined, propagating in the z direction with propagation constant β, and with sinusoidal ($e^{j\omega t}$) time dependence.

On differentiating equation 2.63 *with respect to z* we obtain:

$$\frac{\partial E_x}{\partial z} = -j\beta E_x \qquad (2.64)$$

and on differentiating again:

$$\frac{\partial^2 E_x}{\partial z^2} = -\beta^2 E_x \qquad (2.65)$$

Similarly, on differentiating equation 2.63 twice *with respect to time* we obtain:

$$\frac{\partial^2 E_x}{\partial t^2} = -\omega^2 E_x \qquad (2.66)$$

Substituting 2.65 and 2.66 into 2.62 gives:

$$\frac{\partial^2 E_x}{\partial y^2} = (\beta^2 - \omega^2 \mu_m \varepsilon_m)E_x \qquad (2.67)$$

Since:

$$k_0 = 2\pi/\lambda_0 = \omega/c \tag{2.68}$$

$$1/v^2 = \mu_m \varepsilon_m \tag{2.69}$$

$$c/n = v \tag{2.70}$$

equation 2.67 can be written as:

$$\frac{\partial^2 E_x}{\partial y^2} = (\beta^2 - k_0^2 n_i^2) E_x \tag{2.71a}$$

However, recall that $k_{yi}^2 = k_0^2 n_i^2 - \beta^2$. Hence equation 2.71a can also be written as:

$$\frac{\partial^2 E_x}{\partial y^2} + k_{yi}^2 E_x = 0 \tag{2.71b}$$

Note that n and k_y are now written n_i and k_{yi} because they can represent any of the three media (core, upper cladding, or lower cladding), by letting $i = 1, 2$ or 3.

 Therefore in order to solve for waveguide modes we simply write the solutions of equations 2.71 (one form of the eigenvalue equation) for each of the three media, and apply boundary conditions to determine conditions on the solutions. The generalised solution was described by equation 2.63:

$$E_x = E_x(y)e^{-j\beta z}e^{j\omega t} \tag{2.72}$$

So we use the boundary conditions to find the variation in the y direction (i.e. to find $E_x(y)$). In the upper cladding we will have:

$$E_x(y) = E_u \exp\left[-k_{yu}\left(y - \tfrac{h}{2}\right)\right] \quad \text{for } y \geq (h/2) \tag{2.73}$$

In the core we will have:

$$E_x(y) = E_c \exp[-jk_{yc}y] \quad \text{for } -(h/2) \leq y \leq (h/2) \tag{2.74}$$

In the lower cladding we will have:

$$E_x(y) = E_\ell \exp\left[k_{y\ell}\left(y + \tfrac{h}{2}\right)\right] \quad \text{for } y \leq -(h/2) \tag{2.75}$$

The *boundary conditions* are that both the electric field (E), and its derivative $(\partial E/\partial y)$ are continuous at the boundary. Thus for continuity of the E fields we use equations 2.73–2.75, and substitute for y at the boundaries (i.e. at $y = \pm(h/2)$). Equating 2.73 and 2.74 at $y = h/2$ gives:

$$E_u = E_c \exp\left[-\left(jk_{yc}\frac{h}{2}\right)\right] \tag{2.76}$$

Thus far to retain generality we have expressed the field in the core of the waveguide as an exponential function, representing both sinusoidal and cosinusoidal functions. It is more convenient to change this convention now, in order to visualise the field within the core, and express the field as a sine *or* a cosine function. Modal solutions that are cosine functions are referred to as *even propagation modes* because the cosine is an even function. Similarly modal solutions that are sine functions are referred to as *odd propagation modes*. Since the fundamental mode is an even function, we will proceed with a cosine function, but the process is identical for odd propagation modes. Thus we express the right-hand side of equation 2.76 as a cosinusoidal function, and we can rewrite it as:

$$E_u = E_c \cos\left(k_{yc}\frac{h}{2} + \phi\right) \tag{2.77}$$

Similarly, using the general exponential form, and equating 2.74 and 2.75 at $y = -h/2$, gives:

$$E_\ell = E_c \exp\left[-jk_{yc}\left(-\frac{h}{2}\right)\right] \tag{2.78}$$

As an even function (cosinusoidal function) this can be rewritten as:

$$E_\ell = E_c \cos\left(-k_{yc}\frac{h}{2} + \phi\right) \tag{2.79}$$

We can follow the same process for the derivatives of the E fields $(\partial E/\partial y)$ to obtain:

$$E_u = \frac{k_{yc}}{k_{yu}} E_c \sin\left(k_{yc}\frac{h}{2} + \phi\right) \tag{2.80}$$

$$E_\ell = \frac{k_{yc}}{k_{y\ell}} E_c \sin\left(k_{yc}\frac{h}{2} - \phi\right) \tag{2.81}$$

We now have two expressions for E_u (2.77 and 2.80), and two expressions for E_ℓ (2.79 and 2.81) which we can equate:

$$E_u = E_c \cos\left(k_{yc}\frac{h}{2} + \phi\right) = \frac{k_{yc}}{k_{yu}}E_c \sin\left(k_{yc}\frac{h}{2} + \phi\right) \qquad (2.82)$$

On rearranging, equation 2.82 can be written as:

$$\tan^{-1}\left[\frac{k_{yu}}{k_{yc}}\right] = k_{yc}\frac{h}{2} + \phi + m\pi \qquad (2.83)$$

Similarly from the two expressions for E_ℓ:

$$E_\ell = E_c \cos\left(-k_{yc}\frac{h}{2} + \phi\right) = \frac{k_{yc}}{k_{y\ell}}E_c \sin\left(k_{yc}\frac{h}{2} - \phi\right) \qquad (2.84)$$

So:

$$\tan^{-1}\left[\frac{k_{y\ell}}{k_{yc}}\right] = k_{yc}\frac{h}{2} - \phi + m\pi \qquad (2.85)$$

On adding equations 2.83 and 2.85 we obtain:

$$\tan^{-1}\left[\frac{k_{y\ell}}{k_{yc}}\right] + \tan^{-1}\left[\frac{k_{yu}}{k_{yc}}\right] = k_{yc}h + m\pi \qquad (2.86)$$

Note that the $m\pi$ remains rather than becoming $2m\pi$, because the tangent functions are each periodic in π rather than 2π.

Clearly we can substitute into equation 2.86 for the propagation constants k_i, and we obtain exactly the same eigenvalue equation as before (2.29). This is left as an exercise for the reader. The difference between the two approaches of the ray model and the electromagnetic theory model is that, for the latter, we now have field equations for the modes m, which are a solution of the eigenvalue equation. Before we look at the mode field patterns, let us consider the propagation constants k_i again.

2.7 ANOTHER LOOK AT PROPAGATION CONSTANTS

In solving the wave equation we assumed field solutions with propagation constants in each of the three media (core and upper and lower cladding). However, we did not really discuss why the propagation constants took the form they did. Consider again equations 2.71, which were:

$$\frac{\partial^2 E_x}{\partial y^2} = (\beta^2 - k_0^2 n_i^2)E_x \qquad (2.87a)$$

$$\frac{\partial^2 E_x}{\partial y^2} + k_{yi}^2 E_x = 0 \qquad (2.87b)$$

Our general solution was of the form:

$$E_x = E_c e^{-k_y y} e^{-j\beta z} e^{j\omega t} \qquad (2.88)$$

However, in the core, upper cladding and lower cladding the propagation constant k_y took different forms. In the claddings k_y was a real number, whereas in the core k_y was an imaginary number (which later resulted in a cosine function). This simply corresponds to the condition that total internal reflection is satisfied at both boundaries. Mathematically it corresponds (from 2.87a) to whether β is greater or less than $k_0 n_i^2$. As we know, a term $e^{j\varphi}$ corresponds to a propagating sinusoidal/cosinusoidal type field, whereas $e^{-\varphi}$ simply corresponds to an exponentially decaying field. Thus imaginary propagation constants in the claddings, whilst valid solutions to the wave equation, represent fields propagating in the y direction through the claddings, and hence *not* to a totally internally reflected field. Hence all solutions other than totally internally reflected (guided) waves were ignored.

This does however, provide us with even more insight into the modal solutions of the planar waveguide. Clearly the field penetrates the cladding to a degree determined by the *decay constant* (previously called propagation constant) in the cladding. Hence as the wave propagates, part of the field is propagating in the cladding.

2.8 MODE PROFILES

Now that we have field solutions for modes in the planar waveguide, we can plot the field distribution, $E_x(y)$, or the intensity distribution, $|E_x(y)|^2$. In order to demonstrate more than a single mode, a larger waveguide has been defined than previously, with the following parameters: $n_1 = 3.5, n_2 = 1.5, n_3 = 1.0, \lambda_0 = 1.3\,\mu m$, and $h = 1.0\,\mu m$.

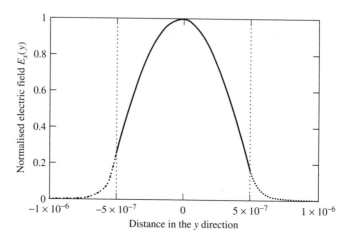

Figure 2.9   Electric field profile of the fundamental mode ($m = 0$)

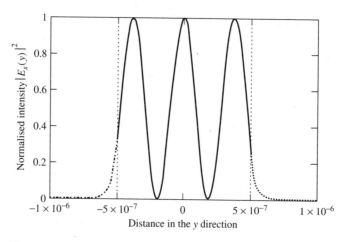

Figure 2.10   Intensity profile of the second even mode ($m = 2$)

For even functions the field in the core is a cosine function, and in the claddings it is an exponential decay. Therefore, solutions for $m = 0$ and $m = 2$ are even functions, and are plotted in Figures 2.9 and 2.10 respectively. The field distribution is plotted for $m = 0$ and the intensity distribution for $m = 2$.

2.9   CONFINEMENT FACTOR

We have seen in the preceding sections that not all of the power propagating in a waveguide mode is contained inside the core of the waveguide. Knowing how much power is inside the core can be useful, and also enables comparisons between waveguide modes and waveguide technologies. Defining a confinement factor is one way of quantifying the confinement. The confinement factor, Γ, is usually defined as expected, as a mathematical expression of the proportion of the power in a given mode that lies within the core:

$$\Gamma = \frac{\displaystyle\int_{-b/2}^{b/2} E_x^2(y)\,dy}{\displaystyle\int_{-\infty}^{\infty} E_x^2(y)\,dy} \tag{2.89}$$

From the preceding section it is clear that modal confinement is a function of polarisation, of the refractive index difference between core and claddings, of the thickness of the waveguide (relative to the wavelength), and of the mode number.

2.10   THE GOOS–HÄNCHEN SHIFT

Let us consider once again the phase change introduced by total internal reflection. On the one hand our ray optic model says that total internal reflection induces a phase change, whereas the electromagnetic model says that the field actually penetrates the cladding layers, and allows us to define the confinement factor. These appear to be conflicting explanations, but a little further development of the ray optic model resolves this apparent ambiguity.

Consider Figure 2.11. We can imagine penetration of both the upper and lower claddings to a depth that is equal to the inverse of the decay constant in the cladding. Clearly this will result in a phase shift of the wave on reflection, and also a lateral shift of the reflected wave, known as the *Goos–Hänchen shift* (shown on the diagram as S_{GH}). By trigonometry, for the upper phase shift we have:

$$\tan\theta_1 = \frac{S_{GH}/2}{1/k_{yu}} = S_{GH}k_{yu}/2 \tag{2.90}$$

so that

$$S_{GH} = \frac{2\tan\theta_1}{k_{yu}} = \frac{2}{k_{yu}}\frac{\beta}{k_{yc}} \tag{2.91}$$

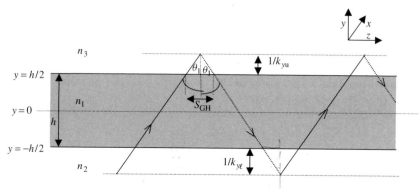

Figure 2.11 Propagation in an asymmetric planar waveguide, showing cladding penetration

This shift can amount to a significant distance, often greater than the waveguide thickness (depending on the waveguide configuration). For example, for the fundamental mode of the waveguide in section 2.8, for which the mode profile was plotted in Figure 2.9, the lateral shift is 0.26 μm. Given that silicon is a strongly confining waveguide system, this is a significant lateral shift. The field outside the core of the waveguide is known as the *evanescent field*. We will see later that this field can be used to practical advantage.

3
Characteristics of Optical Fibres for Communications

3.1 THE STRUCTURE OF OPTICAL FIBRES

Photonic circuits are primarily aimed at applications in communications. As such they usually interface with optical fibres. This chapter is included because it is essential to appreciate the structure of optical fibres in order to understand the difficulties associated with some aspects of optical circuit design. In part the chapter is historical in the sense that it describes characteristics of optical fibres that are now obsolete for some communications applications. However, this approach is deliberate because it clearly demonstrates the origins of some of the well-known characteristics of optical fibres. A good example of this is the discussion of modal dispersion in optical fibres. In long-haul communications, systems now universally use single-mode fibres, and hence modal dispersion is removed. However, unless such a discussion is held, the reader cannot understand the limitations of multimode systems, or why single-mode fibres are now so important.

Optical fibres are available in a variety of forms, each developed for different types of applications. The fibres used in optical communications are usually all silica (SiO_2), with the core and/or the cladding lightly doped to change its refractive index such that the core has a slightly higher refractive index than the cladding. The core may have a constant refractive index, resulting in a step discontinuity in refractive index at the core/cladding interface (the so-called *step-index fibre*), or may have a predetermined refractive index profile that decreases with radial distance from the core centre (*graded-index fibre*). The cladding refractive index

Silicon Photonics: An Introduction   Graham T. Reed and Andrew P. Knights
© 2004 John Wiley & Sons, Ltd   ISBN: 0-470-87034-6

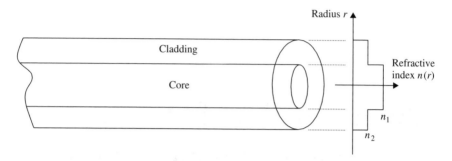

Figure 3.1   The structure of the step-index optical fibre

may be constant or have some specific profile to optimise a particular parametric performance.

The structure of a step-index fibre is shown in Figure 3.1. If the core of the step-index fibre is made sufficiently small, like the planar waveguide it will support a single mode and is referred to as a single-mode fibre. Typically the core of a single-mode fibre will be of the order of $2-10\,\mu m$.

Alternatively the graded-index fibre has a varying refractive index with core radius. Most texts describe the variation as:

$$n(r) = n_1\sqrt{1 - 2\Delta\left(\frac{r}{a}\right)^\alpha} \quad \text{for } r \le a \tag{3.1a}$$

$$n(r) = n_1\sqrt{1 - 2\Delta} = n_2 \quad \text{for } r \ge a \tag{3.1b}$$

where r is the fibre radius, Δ is the maximum relative refractive index difference between the core and the cladding, and α is a parameter which determines the exact profile of the refractive index variation. For optical fibres in which the core and cladding indices are close (the so-called 'weakly guiding approximation'), $\Delta \ll 1$. Communications fibres fall into this category with Δ usually being less than 3 %, and Δ is approximately defined as:

$$\Delta = \frac{n_1 - n_2}{n_1} \tag{3.2}$$

The profiles represented by equations 3.1 are shown in Figure 3.2 for various values of α. It can be demonstrated that, for many applications, $\alpha = 2$ is the optimum (i.e. a parabolic profile), so when reference is made to graded-index fibre a parabolic profile is usually assumed.

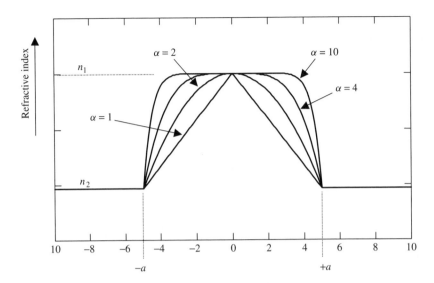

Figure 3.2 Refractive index profile of the graded-index optical fibre core

3.2 MODES OF AN OPTICAL FIBRE

The solution to Maxwell's equations for an optical fibre is much more complex than for the planar waveguide considered in Chapter 2. However, many of the characteristics of guided modes remain. Probably the most important similarity is that the discrete nature of the propagation constant remains, and hence discrete modes also exist. However, the most striking difference between the optical fibre and the planar waveguide is the fact that the optical fibre confines light in two, rather than one, cross-sectional dimension.

The description of optical fibre modes is given in many texts. Here we follow the general approach of Senior [1] because it is straightforward and convenient. Since the optical fibre confines light in two, rather than one, cross-sectional dimension, there is an electric and/or magnetic field variation in two dimensions, which in turn means that two integers, rather than one, are required to describe the modal solution. Whereas planar waveguide modes could be described as TE_0, TE_1 or TM_0, modes of an optical fibre are usually described using integers represented by l and m. The integer l represents the fact that there will be $2l$ field maxima around the circumference of the field distribution, and the integer m refers to the m field maxima along a radius. Hence we refer to TE_{lm} or TM_{lm} modes. However, the propagation characteristics of optical fibres are more complex than simply supporting TE or TM modes. Let us consider the modes of step-index and graded-index fibres in turn.

3.2.1   Modes of a Step-index Fibre

The TE_{lm} or TM_{lm} modes introduced in the last section seem relatively familiar to us following the discussion of modes in the planar waveguide. It requires a small conceptual leap from the planar waveguide to accept these modal solutions of the optical fibre. The guided mode will follow the path of a ray zig-zagging through the fibre, passing through the fibre axis in much the same way as the modes of the planar waveguide propagate. These rays are referred to as *meridional rays*. However, there is another possibility for a ray to propagate through the fibre. This is the so-called *skew ray*, which follows a helical path through the fibre without passing through the fibre axis.

All of these rays are subject to total internal reflection of course, or they would not remain confined within the waveguide, but the propagation angles of the skew rays are such that it is possible for components of both the E and H fields to be transverse to the fibre axis. Hence the modes originating from these rays are designated HE_{lm} or EH_{lm} depending on whether the E or H field dominates the transverse field. However, if we consider only weakly guiding fibres, as mentioned with reference to equation 3.2, the exact modes have very dominant transverse field components, and approximate linearly polarised solutions. This means that several of the modal solutions have approximately the same propagation constant, and therefore the exact modal solutions are usually approximated by linearly polarised modes, designated LP_{lm} modes.

Where two modes that are nominally different modes have the same solution, they are said to be 'degenerate'. By using the LP_{lm} approximation, we make the approximation that several modes become degenerate, because their propagation constants are almost the same. This is demonstrated by Figures 3.3 and 3.4. Figure 3.3 shows the normalised propagation constants of several exact low-order fibre modes. Alternatively Figure 3.4 shows how the exact modes are approximated to LP modes together with the associated intensity profile of the corresponding LP modes. Considering the two diagrams together, we can see that the approximate solutions of Figure 3.4 correspond to normalised propagation constants in Figure 3.3 that are almost equivalent.

To make the approximation even clearer, Table 3.1 shows the correspondence between LP and exact modes.

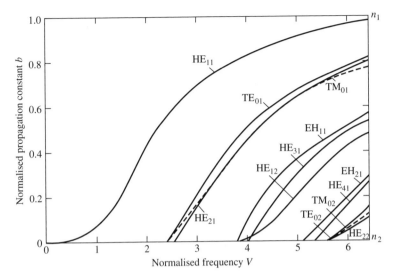

Figure 3.3 Propagation constant of the exact fibre modes plotted against normalised frequency. Reproduced from G P Agrawal (1997) *Fiber Optic Communications Systems*, 2nd edn, John Wiley & Sons, New York by permission of John Wiley and Sons Inc.

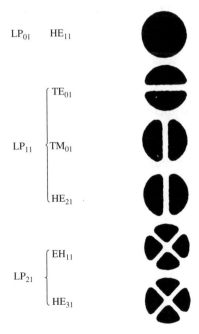

Figure 3.4 Intensity profiles of the three lowest-order LP modes. Source: *Optical Fiber Communications, Principles and Practice*, 2nd edn, J M Senior, Pearson Education Limited. Reproduced by permission of Pearson Education

Table 3.1 Relationship between approximate LP modes and exact modes. Source: *Optical Fiber Communications, Principles and Practice*, 2nd edn, J M Senior, Pearson Education Limited. Reproduced by permission of Pearson Education

Linearly polarised	Exact modes
LP_{01}	HE_{11}
LP_{11}	$HE_{21}, TE_{01}, TM_{01}$
LP_{21}	HE_{31}, EH_{11}
LP_{02}	HE_{12}
LP_{31}	HE_{41}, EH_{21}
LP_{12}	$HE_{22}, TE_{02}, TM_{02}$
LP_{lm}	$HE_{2m}, TE_{0m}, TM_{0m}$
$LP_{lm} (l \neq 0 \text{ or } 1)$	$HE_{l+1,m}, EH_{l-1,m}$

3.2.2　Modes of a Graded-index Fibre

Both meridional and skew mode solutions exist also in the graded-index fibre, but the most striking difference between step-index and graded-index fibres is the path the meridional rays take when propagating along the fibre. A simple way to visualise this distinction is given by Senior [1], which we will follow here. Consider a graded-index fibre with a parabolic refractive index profile (i.e. $\alpha = 2$), as shown schematically in Figure 3.5.

Since the refractive index varies radially, we can imagine the variation to be a series of discrete steps in refractive index, as shown in Figure 3.6. We can then imagine a ray being refracted at each interface, thus having an increasing angle of incidence with increasing radius. Eventually the ray will be totally internally reflected at one of the imaginary interfaces, and will propagate back towards the fibre axis. Thus the ray may never reach the core/cladding interface before being totally internally reflected. In reality, the continuum of changing refractive index will bend the

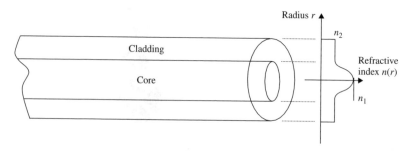

Figure 3.5　Structure of the graded-index optical fibre

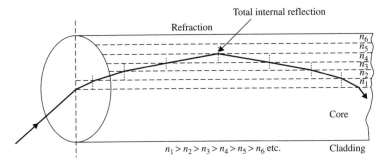

Figure 3.6 Total internal reflection in a graded-index fibre. Source: *Optical Fiber Communications, Principles and Practice*, 2nd edn, J M Senior, Pearson Education Limited. Reproduced by permission of Pearson Education

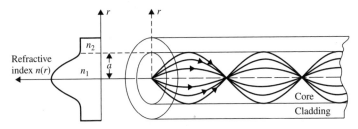

Figure 3.7 Mode trajectories in graded-index fibre. Source: *Optical Fiber Communications, Principles and Practice*, 2nd edn, J M Senior, Pearson Education Limited. Reproduced by permission of Pearson Education

ray rather than it undergoing a series of discrete refractions, but the net result is similar. In this way, we can imagine a series of different modes following different paths down the fibre, each reaching a different maximum radius within the fibre, as shown in Figure 3.7.

3.3 NUMERICAL APERTURE AND ACCEPTANCE ANGLE

We have seen in preceding sections that modes of an optical waveguide propagate with specific propagation constants or mode angles. When coupling light to an optical fibre, there is an additional boundary between two media of different refractive indices, usually that of the core and air. In order to couple light into modes that will be totally internally reflected within the waveguide, a range of angles exist outside the waveguide that correspond to these waveguide modes. This situation is illustrated in Figure 3.8, in which a multimode, step-index fibre is assumed.

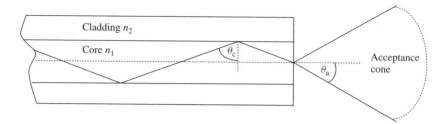

Figure 3.8   Acceptance angle of an optical fibre

The mode depicted within the fibre represents propagation at the critical angle θ_c, which is translated to an acceptance cone of angles outside the fibre, with acceptance angle θ_a. Any ray that enters the fibre at an angle to the fibre axis of greater than θ_a will not be totally internally reflected at the core/cladding interface. The acceptance angle translates to an acceptance cone owing to the circular symmetry of the fibre.

We can take this ray analysis a little further by considering Snell's law at the fibre core/air interface, and at the core/cladding interface. At the core/air interface:

$$n_a \sin \theta_a = n_1 \sin[90° - \theta_c]$$

$$= n_1 \cos \theta_c$$

$$= n_1 \sqrt{(1 - \sin^2 \theta_c)} \qquad (3.3)$$

since $\sin^2 \theta + \cos^2 \theta = 1$. But we recall that the critical angle is defined as:

$$\sin \theta_c = \frac{n_2}{n_1} \qquad (3.4)$$

Hence equation 3.3 becomes:

$$n_a \sin \theta_a = n_1 \sqrt{(1 - n_2^2/n_1^2)} = \sqrt{n_1^2 - n_2^2} \qquad (3.5)$$

Equation 3.5 relates the fibre acceptance angle to the refractive indices of the core and cladding. This equation also forms the basis of an important optical fibre parameter known as the *numerical aperture* (*NA*). Thus the *NA* is defined as:

$$NA = n_a \sin \theta_a = \sqrt{n_1^2 - n_2^2} \qquad (3.6)$$

Since the numerical aperture is usually defined in air, the value of n_a is usually 1.0, and $NA = \sin\theta_a$. We can also define NA in terms of the relative refractive index Δ, using equation 3.2:

$$NA = n_1\sqrt{\Delta} \qquad (3.7)$$

Note that equation 3.7 is sometimes written as $NA = n_1\sqrt{2\Delta}$ owing to a slightly different definition of Δ.

We can see that the numerical aperture is independent of core diameter, and is a useful measure of the ability of the fibre to receive light. However, the definition has been derived using only meridional rays, and is not valid for very small fibre diameters where the ray model in not accurate, and electromagnetic theory must be used. A similar analysis can be carried out for skew rays (see, for example, Chapter 2 in Senior [1]). The resulting expression for NA is:

$$NA = n_a \sin\theta_a \cos\gamma = \sqrt{n_1^2 - n_2^2} \qquad (3.8)$$

where γ is the angle between the plane of total internal reflection for the skew rays, and the projection of the ray onto a plane normal to the fibre axis (see Figure 3.9 for clarification).

Comparison of equations 3.6 and 3.8 shows that skew rays are accepted at larger angles than meridional rays, by virtue of the term '$\cos\gamma$', hence effectively extending the acceptance cone of the fibre, and increasing the light gathering capacity. However, for communications purposes the definition of equation 3.9 is usually considered adequate.

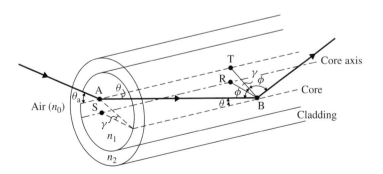

Figure 3.9 The acceptance angle of skew rays. Source: *Optical Fiber Communications, Principles and Practice*, 2nd edn, J M Senior, Pearson Education Limited. Reproduced by permission of Pearson Education

3.4   DISPERSION IN OPTICAL FIBRES

Dispersion in optical fibres means that parts of the signal propagate through the fibre at slightly different velocities, resulting in distortion of the signal. Whilst dispersion affects both analogue and digital signals, it is most easily demonstrated by considering a digital signal.

The effect on a digital signal will be for a pulse that may have been developed at the transmission point of the fibre as a near square wave, degenerating in shape to something resembling a gaussian function. With increasing propagation distance, the pulse will broaden and begin to overlap with the next pulse in the pulse chain, until eventually it will no longer be possible to distinguish between successive pulses in the pulse chain.

It is possible to estimate the bit rate possible in a fibre with a given pulse broadening, $\delta\tau$. If pulses are not to overlap, then the maximum pulse broadening must be a maximum of half of the transmission period, T (which is often twice the pulse length):

$$\delta\tau \leq \tfrac{1}{2}T \tag{3.9}$$

where the bit rate, B_T, is $1/T$. Hence the maximum bit rate is:

$$B_T \leq \frac{1}{2\delta\tau} \tag{3.10}$$

Some texts use a more conservative rule of thumb, making $B_T \leq 1/4\delta\tau$. However, perhaps a more realistic and hence accurate estimate is given by considering the pulses to be gaussian, with an r.m.s. width σ. Because each pulse is no longer considered to be rectangular, a certain amount of overlap is possible whilst still distinguishing between the pulses. A typical rule of thumb is $B_T \leq 0.2/\sigma$ [3].

The dispersion itself is a result of a number of factors, combined here under the categories of either intermodal dispersion or intramodal dispersion.

3.4.1   Intermodal Dispersion

Intermodal dispersion is a result of the different propagation times of different modes within a fibre. Clearly the maximum difference in propagation times is between the fastest and slowest modes. This is most easily demonstrated by considering a multimode step-index fibre.

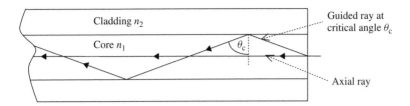

Figure 3.10   The origin of modal dispersion

The fastest and slowest modes possible in such a fibre will be the axial ray, and the ray propagating at the critical angle θ_c, as depicted in Figure 3.10. Since the two rays are travelling at the same speed within the core of the fibre (i.e. constant refractive index), the differential delay is primarily related to the different path lengths of the two rays. Thus the time for the axial ray to travel the length L of the fibre is:

$$t_{min} = \frac{distance}{velocity} = \frac{L}{c/n_1} = \frac{Ln_1}{c} \tag{3.11}$$

Similarly for the slowest ray (Figure 3.10), the maximum time, is:

$$t_{max} = \frac{L/\sin\theta_c}{c/n_1} = \frac{L}{c}\frac{n_1}{\sin\theta_c} \tag{3.12}$$

Furthermore, recalling from equation 3.4 that the critical angle is $\sin\theta_c = \frac{n_2}{n_1}$, equation 3.12 becomes:

$$t_{max} = \frac{L/\sin\theta_c}{c/n_1} = \frac{L\,n_1^2}{c\,n_2} \tag{3.13}$$

Therefore the delay difference between the fastest and slowest modes is:

$$\delta t_{si} = t_{max} - t_{min} = \frac{L\,n_1^2}{c\,n_2} - \frac{Ln_1}{c}$$

$$= \frac{L\,n_1^2}{c\,n_2}\left(\frac{n_1 - n_2}{n_1}\right) = \frac{L\,n_1^2}{c\,n_2}\Delta \tag{3.14}$$

where Δ is the relative refractive index. The subscript 'si' indicates that this time delay difference is associated with the step-index fibre. Since we have assumed the weakly guiding approximation, $n_1 \approx n_2$, and hence equation 3.14 is often written as:

$$\delta t_{\text{si}} = \frac{L n_1 \Delta}{c} \tag{3.15}$$

Note that this simple analysis has ignored skew rays. However, it gives us a reasonable idea of the relative delay of the fastest and slowest modes. We can also use equation 3.10 to estimate the maximum bit rate for a step-index multimode fibre. To do this we need to assume some representative values of Δ and n_1. If we let $\Delta = 0.01$ and $n_1 = 1.49$. Hence from equation 3.14, $\delta \tau = 49.6 \, \text{ns/km}$. Using equation 3.10, the maximum bit rate becomes:

$$B_T = \frac{1}{2 \delta \tau} \approx 10 \, \text{MBit/s·km} \tag{3.16}$$

This restrictive bandwidth is improved significantly by using graded-index fibre. We have seen that the trajectory of rays in a graded-index fibre changes with propagation angle in the same way as in a step-index fibre. However, the grading of the refractive index means that, since the velocity of the ray is inversely proportional to the local refractive index, the modes which travel significantly in the regions of lower refractive index travel faster during these periods. Hence the longer paths travelled by modes with greater propagation angles is somewhat compensated by higher propagation velocity in the lower refractive indices at the extremes of the fibre core. Therefore there is an equalisation of transit times for the different rays in a graded-index fibre. The resultant difference in time delay can be written as [1]:

$$\delta t_{\text{gi}} = \frac{L n_1 \Delta^2}{8c} \tag{3.17}$$

For comparison purposes, if we again let $\Delta = 0.01$ and $n_1 = 1.49$, equation 3.17 gives $\delta \tau = 62 \, \text{ps/km}$. Using equation 3.10, the maximum bit rate becomes:

$$B_T = \frac{1}{2 \delta \tau} \approx 8 \, \text{GBit/s·km} \tag{3.18}$$

Clearly this is a significant improvement over the 10 Mbit/s·km for the step-index fibre. However, this is not sufficient for long-haul communications, and so for such applications modal dispersion is eliminated by using single-mode fibre.

3.4.2 Intramodal Dispersion

Intramodal dispersion results from the fact that any optical source is not purely monochromatic, so the different spectral components of the light source may have different propagation delays, and hence pulse broadening may occur, even if transmitted by a single mode. Of course the phenomenon is also present in multimode fibres, but it is usually dominated by intermodal dispersion. Since it is due to spectral variation of the source, the term 'chromatic dispersion' is used interchangeably with the term 'intramodal dispersion'.

In order to understand chromatic dispersion, it is first necessary to be precise about another parameter, the *velocity* of a pulse in the fibre. It is well known that a pulse consisting of a finite spread of wavelengths will propagate with a group velocity, which will, in general, be different from the phase velocity of the wavefront. (For a discussion of phase and group velocity, see for example Kazovsky et al. [4].) Without explicitly stating it, our discussions so far have generally assumed truly monochromatic light, and when we have discussed the velocity of light in a waveguide we have implicitly used the phase velocity, defined as:

$$v_p = \omega/\beta \qquad (3.19)$$

However, the group velocity is defined as:

$$v_g = \delta\omega/\delta\beta \qquad (3.20)$$

Therefore the propagation delay at any given value of frequency, ω (and hence wavelength λ), is given by:

$$\tau = L/v_g \qquad (3.21)$$

Differentiating equation 3.21 will allow us to find the variation in propagation delay with frequency (wavelength):

$$\frac{\partial \tau}{\partial \omega} = L \frac{\partial}{\partial \omega}\left(\frac{1}{v_g}\right) \qquad (3.22)$$

Using equation 3.20 to substitute for v_g, equation 3.22 becomes:

$$\frac{\partial \tau}{\partial \omega} = L \frac{\partial^2 \beta}{\partial \omega^2} \qquad (3.23)$$

Therefore, for an optical source of spectral width $\Delta\omega$, the difference in propagation delay between components at either extreme of this spectral width will be:

$$\delta\tau_{ch} = \left|\frac{\partial\tau}{\partial\omega}\right|\Delta\omega = \left|\frac{\partial^2\beta}{\partial\omega^2}\right|L\Delta\omega \qquad (3.24)$$

In equation 3.24 the subscript 'ch' denotes *chromatic dispersion*. Chromatic dispersion is usually described using the so-called group velocity dispersion (GVD) parameter, denoted D. The parameter D is related to the term $\partial^2\beta/\partial\omega^2$ by the expression:

$$D = -\frac{2\pi c}{\lambda^2}\left(\frac{\partial^2\beta}{\partial\omega^2}\right) \qquad (3.25)$$

D is expressed in units of ps/(km·nm). Therefore equation 3.24 can be expressed in terms of D as:

$$\delta\tau_{ch} = |D|L\Delta\lambda \qquad (3.26)$$

There are two main contributions to chromatic dispersion:

1   *Material dispersion.* The refractive index of any medium is a function of wavelength, and hence different wavelengths that see different refractive indices will propagate with different velocities, resulting in intramodal dispersion.

2   *Waveguide dispersion.* Even if the refractive index is constant, and material dispersion eliminated, the propagation constant β would vary with wavelength for any waveguide structure, resulting in intramodal dispersion.

It is possible to separate these contributions to chromatic dispersion (see for example Agrawal [2]), although there is little advantage for the purposes of this text. It is instructive, however, to observe the variation in chromatic dispersion with wavelength, for a single-mode fibre. Figure 3.11 shows total chromatic dispersion D, together with the contributions from material dispersion D_M, and from waveguide dispersion D_W. It is interesting to note that there is zero dispersion close to 1.31 µm, where the curve passes through zero, marked on the diagram as λ_{ZD}. It is also worth noting that the principal effect of waveguide dispersion is to shift the zero dispersion wavelength λ_{ZD} along the wavelength axis.

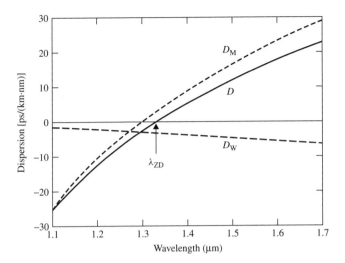

Figure 3.11 The variation in chromatic dispersion with wavelength. Reproduced from G P Agrawal (1997) *Fiber Optic Communications Systems*, 2nd edn, John Wiley & Sons, New York by permission of John Wiley and Sons Inc.

Since the contribution of waveguide dispersion is dependent on fibre parameters such as refractive indices and core diameter, it is possible to vary the contribution, to shift λ_{ZD} typically to 1.55 μm, where the optical loss of optical fibre is a minimum. Alternatively the contribution of waveguide dispersion can be varied in such a way as to make total chromatic dispersion relatively small over the wavelength range 1.3–1.6 μm. Such fibres are called *dispersion shifted fibres* and *dispersion flattened fibres* respectively, and their dispersion characteristics are shown in Figure 3.12.

It can be seen from Figures 3.11 and 3.12 that the total chromatic dispersion is of the order of a few ps/(nm·km). In order to make a comparison between single-mode fibres and multimode fibres, let us assume that we operate at a wavelength within a few nanometers of λ_{ZD}, and hence a figure of 1 ps/(nm·km) would not be unreasonable for D. Let us further assume a spectral width for the light source of 1 nm. Then from equation 3.26 the pulse broadening $\delta\tau_{ch}$ is approximately 1 ps per kilometer. Using this value in equation 3.10 we can estimate the maximum bit rate for single-mode fibre:

$$B_T = \frac{1}{2\delta\tau_{ch}} \approx 500 \, \text{GBit/s·km} \tag{3.27}$$

Clearly this is a much larger value than for multimode fibre.

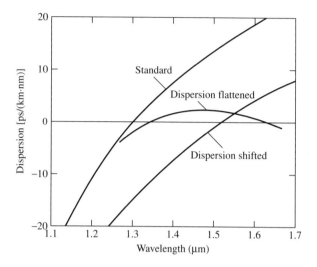

Figure 3.12 Dispersion-shifted and dispersion-flattened fibres. Reproduced from G P Agrawal (1997) *Fiber Optic Communications Systems*, 2nd edn, John Wiley & Sons, New York by permission of John Wiley & Sons Inc.

3.5 SINGLE-MODE FIBRES: MODE PROFILE, MODE-FIELD DIAMETER, AND SPOT SIZE

So far we have discussed the modes of optical fibres without considering the electric field distribution of the modes explicitly, in the way we did for planar waveguides in Chapter 2. In this section we will briefly consider the electric field profile for a single-mode step-index fibre. An optical fibre is cylindrical in form, so when solving the wave equation for optical fibres cylindrical coordinates $(r, \phi$ and $z)$ are used, rather than cartesian coordinates $(x, y$ and $z)$. The solution of the wave equation for the fibre results in solutions of the form 'exp $j\beta z$' in the z direction, 'exp $jm\phi$' around the circumference of the fibre (where m must be an integer to ensure periodic solutions with period 2π), and radial solutions which are Bessel functions in the core and modified Bessel functions in the cladding. Since Bessel functions are damped oscillatory functions, and modified Bessel functions decay to zero at infinite distance, the resultant radial field solutions resemble the cross-sectional field variations of the planar waveguide discussed in Chapter 2. For information, Bessel functions and modified Bessel functions are reproduced in Figure 3.13.

For a single-mode fibre, only the fundamental mode is supported, and its radial field distribution takes the form of a Bessel function in the core, $J_0(r)$, matched to modified Bessel functions in the cladding, $K_0(r)$. This function is shown in Figure 3.14 for two different values

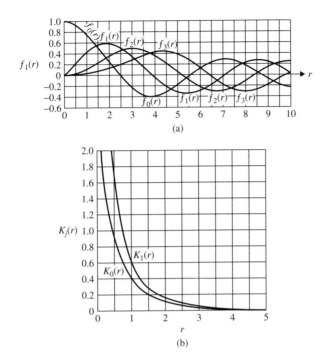

Figure 3.13 (a) Variation of the first four Bessel functions. (b) Variation of the first two modified Bessel functions. Source: *Optical Fiber Communications, Principles and Practice*, 2nd edn, J M Senior, Pearson Education Limited. Reproduced by permission of Pearson Education

of normalised frequency (discussed later). We can note from this figure that a significant amount of the propagating power is transmitted in the cladding. Therefore the diameter of the cladding is sometimes less important than the extent to which the field penetrates. A parameter that is used to represent the extent to which the field penetrates is known as the *mode-field diameter* (MFD). A variety of definitions exist for mode-field diameter, but one of the simplest forms uses an approximation to the true mode-field distribution. Consideration of Figure 3.14 reveals that the shape of the field profile is very similar to a gaussian function.

Consequently a gaussian function is often used to approximate the fundamental field profile, and for the purposes of defining MFD a very convenient definition follows from this approximation. In this case the mode-field diameter is taken as the $1/e$ width of the function; i.e. the width of the function at an amplitude of $1/e$ (≈ 0.37) of the peak value. This corresponds to $1/e^2$ of the power distribution. The MFD is shown diagrammatically in Figure 3.15. Another parameter, the spot size (or mode-field radius), ω_0, is also shown in the diagram.

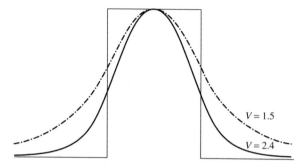

Figure 3.14 Field shape of the fundamental mode of a step-index fibre, for normalised frequencies of $V = 1.5$ and $V = 2.4$. Source: *Optical Fiber Communications, Principles and Practice*, 2nd edn, J M Senior, Pearson Education Limited. Reproduced by permission of Pearson Education

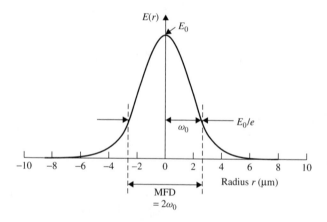

Figure 3.15 Approximation to the fundamental mode, showing the mode-field diameter (MFD) and the spot size ω_0. Source: *Optical Fiber Communications, Principles and Practice*, 2nd edn, J M Senior, Pearson Education Limited. Reproduced by permission of Pearson Education

3.6 NORMALISED FREQUENCY, NORMALISED PROPAGATION CONSTANT, AND CUTOFF WAVELENGTH

Thus far the parameters normalised frequency and normalised propagation constant have not been defined explicitly, although they have appeared in several diagrams. This has been deliberate, in order to avoid unnecessary complications in the discussions of some introductory concepts. However, since these parameters are used extensively in the literature, they are now defined. The normalised frequency, V, is defined as:

$$V = \frac{2\pi}{\lambda_0} a \sqrt{n_1^2 - n_2^2} = \frac{2\pi}{\lambda_0} a n_1 \sqrt{2\Delta} \qquad (3.28)$$

The normalised frequency is sometimes referred to simply as the V number, and it is clearly related to a specific fibre design, as it contains the fibre core radius, a, the relative refractive index difference, Δ, and the operating wavelength, λ_0. It is a very useful parameter, as it allows general solutions to be found for all optical fibres, if the general solution is written in terms of the V number. The V number of a particular fibre can then be used to interpret the general solution.

Similarly, the normalised propagation constant can be defined as:

$$b = \frac{(\beta/k_0)^2 - n_2^2}{n_1^2 - n_2^2} = \frac{(\beta/k_0)^2 - n_2^2}{2n_1^2\Delta} \qquad (3.29)$$

Since β must lie between the limits of $n_2 k_0$ and $n_1 k_0$, b lies between 0 and 1.

Figure 3.3 showed the variation of the normalised propagation constant with the normalised frequency. It is reproduced here as Figure 3.16

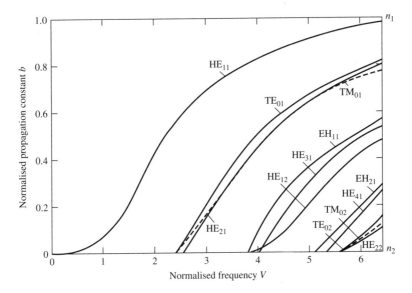

Figure 3.16 Propagation constant of the exact fibre modes plotted against normalised frequency. Reproduced from G P Agrawal (1997) *Fiber Optic Communications Systems*, 2nd edn, John Wiley & Sons, New York by permission of John Wiley and Sons Inc.

for convenience. It can be seen from this diagram that there is value of normalised frequency below which only the HE_{11} mode propagates, and is therefore single mode. Therefore this value is termed the *cutoff normalised frequency*, V_c. It is 2.405 for step-index fibres.

We can rearrange equation 3.28 in terms of wavelength, in order to express the cutoff wavelength explicitly:

$$\lambda_c = \frac{2\pi}{V_c} an_1 \sqrt{2\Delta} \tag{3.30}$$

The expression can be simplified even further by dividing 3.30 by 3.28, to obtain $\lambda_c/\lambda_0 = V/V_c$. Since $V_c = 2.405$ for step-index fibres, then:

$$\lambda_c = \frac{V\lambda_0}{2.405} \tag{3.31}$$

Therefore, using equation 3.31, and knowing the required operating wavelength, the cutoff frequency for a fibre with a given V number can be found easily.

REFERENCES

1. J M Senior (1992) *Optical Fiber Communications, Principles and Practice*, 2nd edn, Prentice-Hall, London.
2. G P Agrawal (1997) *Fiber Optic Communications Systems*, 2nd edn, John Wiley & Sons, New York.
3. W A Gambling, A H Hartog and C M Ragdale (1981) 'Optical fibre transmission lines', *Radio Electron. Eng.* (IERE), 51, 313–325.
4. L Kazovsky, S Benedetto and A Willner (1996) *Optical Fiber Communication Systems*, Artech House, Boston.

Silicon-on-Insulator (SOI) Photonics

4.1 INTRODUCTION

Integrated optics in silicon is interesting for a combination of techno-logical and cost reasons. The cost issues are relatively obvious, and are related to the absolute cost of silicon and SOI wafers as compared to more exotic materials such as the III–V compounds or the insulator lithium niobate ($LiNbO_3$), to the cost of processing the wafers, and to the packing density that can be achieved. Silicon is a well-understood and robust material, and the processing of silicon has been developed by the electronics industry to a level that is more than sufficient for most integrated optical applications. The minimum feature size for most applications is currently of the order of 1–2 microns, which in terms of microelectronics is very old technology, although there is a trend towards miniaturisation. Furthermore, as silicon microelectronics con-tinues to advance, new and improved processing becomes available. One exception to this generalisation is the occasional requirement for a very compact structure in silicon, such as a grating, which, owing to the large refractive index of silicon ($\sim$3.5), typically needs to have a submicron minimum feature size for applications at wavelengths at which silicon is transparent ($>1.1\,\mu$m approx.). It could be argued that an alternative technology, that of silica-based integrated optics, is lower cost, but this is a passive material with little prospect for active devices such as sources, detectors, or optical modulators (other than thermally operated) becoming available. Consequently this technological issue is intimately linked to cost.

Silicon Photonics: An Introduction   Graham T. Reed and Andrew P. Knights
© 2004 John Wiley & Sons, Ltd   ISBN: 0-470-87034-6

One significant technological issue is associated with the possibility of optical phase and amplitude modulation in silicon. Refractive index modulation is possible via several techniques, although it is now widely accepted that free carrier injection is the most efficient of these. This is not a fast modulation mechanism when compared to the field-effect mechanisms available in other technologies, dominated by the linear electro-optic (Pockels) effect, used to advantage in, for example, lithium niobate at operating speeds of several tens of gigahertz. However, carrier injection is sufficient for many communications and sensor applications. Currently carrier injection modulators in SOI are limited to modulation bandwidths of the order of a few tens of megahertz, although this is not a fundamental limit and it will be discussed later in more detail. The Pockels effect is not observed in silicon owing to the centro-symmetric nature of the crystal structure.

Sources and detectors (beyond $1.1\,\mu m$) in silicon have not been reported with sufficient efficiency to make them commercially viable as yet, but this is a current active research topic for many groups around the world. The topic of light sources in silicon is a particularly active research activity at present, and is discussed further in Chapter 8. Nevertheless the possibility of accurate micromachining of silicon has meant that sources and detectors can be accurately aligned to silicon waveguides in a hybrid circuit, and silicon sources and detectors will need to be very cost-effective to displace this approach, or offer very significant improvements in performance.

4.2   SILICON-ON-INSULATOR WAVEGUIDES

Thus far, when discussing optical waveguides other than optical fibres, the discussion has been limited to a theoretical three-layer structure comprising a core and two cladding layers, the latter infinite in the x and y directions. Clearly this is not a practical structure. However, the silicon-on-insulator planar waveguide approximates this theoretical waveguide.

The configuration of an SOI wafer is shown in Figure 4.1, and the fabrication of such wafers is discussed in Chapter 5.

The silicon guiding layer is typically of the order of a few micrometers in thickness, and the buried silicon dioxide is typically about half a micron, although individual designs vary. The purpose of the buried oxide layer is to act as the lower cladding layer, and hence prevent the field associated with the optical modes from penetrating the silicon substrate below. Therefore as long as the oxide is thicker than the

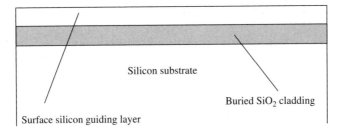

Figure 4.1 SOI planar waveguide

evanescent fields associated with the modes it will be satisfactory. Sometimes a surface oxide layer is also introduced as a passivation layer. If we consider the three-layer model of Chapter 2, the waveguide is transformed from an asymmetrical waveguide into a symmetrical one by the addition of a surface oxide layer. In practice, however, the refractive indices of both air ($n = 1$) and silicon dioxide ($n \approx 1.5$) are so different from that of silicon ($n \approx 3.5$) that the two configurations are very similar (see Chapter 2).

Whilst the planar waveguide is convenient as an introduction to the subject of optical waveguides, it is of limited practical use, because light is confined only in one dimension. For many applications, two-dimensional confinement is required. This is achieved in silicon by etching a two-dimensional waveguide, the most common being the so-called 'rib' waveguide. The cross-section of a rib waveguide is shown in Figure 4.2.

Clearly this is just one way to achieve optical confinement in two dimensions, but based on what we have studied so far in this text, it is perhaps not obvious why the rib structure achieves optical confinement in the lateral direction. This is discussed further below.

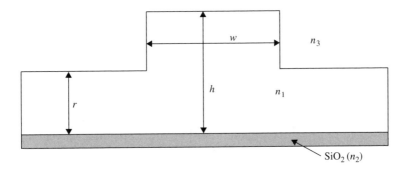

Figure 4.2 Rib waveguide geometry

4.2.1 Modes of Two-dimensional Waveguides

In common with modes of optical fibres, two-dimensional rectangular waveguide modes require two subscripts to identify them. However, it is not appropriate to use the same convention as used in optical fibres, because rectangular guides are clearly not circularly symmetric. Therefore it is natural to use cartesian coordinates. Unfortunately two slightly different conventions have developed, although they are very similar. In a rectangular waveguide there exist two families of modes, the HE mode and the EH modes. In common with the skew modes of an optical fibre, these are mostly polarised in the TE or TM directions. Therefore these modes are usually referred to as the E^x or E^y modes depending on whether they are mostly polarised in the x or y direction. Two subscripts are then introduced to identify the mode. Therefore the modes are designated $E^x_{p,q}$ or $E^y_{p,q}$, where the integers p and q represent the number of field maxima in the x and y directions respectively. These modes are also referred to as $HE_{p,q}$ and $EH_{p,q}$ modes. Hence the fundamental modes are referred to as $E^x_{1,1}$ and $E^y_{1,1}$. This is probably the most common convention.

However, a second convention also exists in which the form of mode identification is essentially the same, but the integers start at 0 rather than at 1. This has developed from the convention used for planar waveguides used in Chapter 2, in which the fundamental mode is the mode for which the integer $m = 0$. Hence in the second convention for rectangular waveguides, the fundamental modes are referred to as $E^x_{0,0}$ and $E^y_{0,0}$ (or $HE_{0,0}$ and $EH_{0,0}$). Hence some care must be taken to indicate the labelling convention used.

4.3 THE EFFECTIVE INDEX METHOD OF ANALYSIS

In Chapter 2 the effective index of an optical mode was introduced. In this section the effective index is used to find approximate solutions for the propagation constants of two-dimensional waveguides. This is done without resorting to any discussion of electric fields within the waveguide, and so the simplicity of the method, known as the effective index method, is very attractive.

The method will be first discussed in general terms, before an example is carried out. Consider a generalised two-dimensional waveguide, as shown in Figure 4.3a. The approach to finding the propagation constants for the waveguide shown is to regard it as a combination of two planar waveguides, one horizontal and one vertical. We then successively

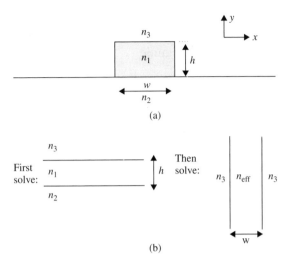

Figure 4.3 (a) Generalised two-dimensional waveguide. (b) Decomposition into two imaginary planar waveguides

solve the planar waveguide eigenvalue equations first in one direction and then the other, taking the effective index of the first as the core refractive index for the second. Care must be taken to consider the polarisation involved. If we are considering an electric field polarised in the x direction (TE polarisation), then when solving the three-layer planar waveguide in the y direction we use the TE eigenvalue equation. However, when we subsequently solve the vertical three-layer planar waveguide, we must use the TM eigenvalue equation because, with respect to this imaginary vertical waveguide, the field is polarised in the TM direction. Decomposition of the two-dimensional waveguide into two planar waveguides is depicted in Figure 4.3b.

The situation is complicated when solving a rib waveguide structure, because the refractive index on either side of the core is not constant over the height of the core. Therefore, when solving the first part of the decomposed two-dimensional waveguide, we need to find the effective index, not only in the core, but also in the slab regions at each side of the core in the x direction, prior to solving the second part of the decomposition. This is best illustrated by an example.

Example

Find the effective index, N_{wg}, of the fundamental mode of the rib waveguide in Figure 4.2, for the following waveguide parameters:

$w = 3.5\,\mu\text{m}$, $h = 5\,\mu\text{m}$, $r = 3\,\mu\text{m}$, $n_1 = 3.5$, $n_2 = 1.5$, $n_3 = 1.0$. The operating wavelength is $1.3\,\mu\text{m}$, and TE polarisation is assumed. This is representative of a silicon waveguide.

Solution

Firstly we must decompose the rib structure into vertical and horizontal planar waveguides, as shown in Figure 4.3b. This means we must first solve the horizontal planar waveguide, shown in Figure 4.4. Since the polarisation is TE, we use the asymmetrical TE eigenvalue equation developed in Chapter 2. This was equation 2.30, reproduced here as equation 4.1:

$$[k_0 n_1 h \cos\theta_1 - m\pi] = \tan^{-1}\left[\frac{\sqrt{\sin^2\theta_1 - (n_2/n_1)^2}}{\cos\theta_1}\right]$$
$$+ \tan^{-1}\left[\frac{\sqrt{\sin^2\theta_1 - (n_3/n_1)^2}}{\cos\theta_1}\right] \qquad (4.1)$$

Solving equation 4.1 for the conditions above and $m = 0$ yields a propagation angle of $87.92°$ (1.53456 radians). From equation 2.21:

$$\beta = n_1 k_0 \sin\theta_1 \qquad (4.2)$$

Therefore the effective index of the waveguide region is given by:

$$n_{\text{effg}} = \frac{\beta}{k_0} = n_1 \sin\theta_1 = 3.4977. \qquad (4.3)$$

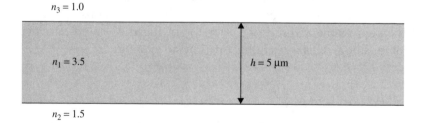

$n_3 = 1.0$

$n_1 = 3.5$

$h = 5\,\mu\text{m}$

$n_2 = 1.5$

Figure 4.4   First planar waveguide of the decomposed rib structure of Figure 4.2

We now need to solve the second decomposed planar waveguide of the decomposed rib structure. However, in order to do so, we must first find the effective index of the planar regions each side of the core. To do this we must solve the asymmetrical TE eigenvalue equation again, this time for a waveguide height of $r = 3\,\mu\text{m}$. This yields a propagation angle of $86.595°$, and an effective index for the planar region of $n_{\text{effp}} = 3.4938$.

We can now solve the second decomposed planar waveguide of the decomposed rib structure, using the effective indices just calculated. This planar waveguide structure is shown in Figure 4.5. This time the TM eigenvalue equation should be used, and note that now we also have a symmetrical waveguide. The symmetrical TM eigenvalue equation for the fundamental mode is (from equation 2.27):

$$k_0 n_{\text{effg}} w \cos\theta_{\text{wg}} = 2\tan^{-1}\left[\frac{\sqrt{\left(\dfrac{n_{\text{effg}}}{n_{\text{effp}}}\right)^2 \sin^2\theta_{\text{wg}} - 1}}{\left(\dfrac{n_{\text{effp}}}{n_{\text{effg}}}\right)\cos\theta_{\text{wg}}}\right] \tag{4.4}$$

where θ_{wg} is be found. The solution is $\theta_{\text{wg}} = 88.285°$, which corresponds to an effective index of $N_{\text{wg}} = 3.496$. Knowing the effective index we can evaluate all of the propagation constants in the core and cladding, using the equations in Chapter 2. In particular we are interested in the z-directed propagation constant:

$$\beta = k_0 N_{\text{wg}} = 16.897\,\text{rad}\cdot\mu\text{m}^{-1} \tag{4.5}$$

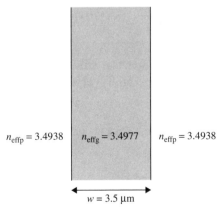

$n_{\text{effp}} = 3.4938$ $n_{\text{effg}} = 3.4977$ $n_{\text{effp}} = 3.4938$

$w = 3.5\,\mu\text{m}$

Figure 4.5   Second planar waveguide of the decomposed rib structure of Figure 4.2

There are several things of note from the above example. Firstly the effective indices are all very close to the refractive index of silicon that was assumed initially, of $n = 3.5$. This is expected because we have used a large rib waveguide as our example, and hence the fundamental mode will be well confined – most of the power in the fundamental mode propagates within the silicon layer itself, with glancing propagation angles (according to the ray model!). Secondly the confinement in the vertical and horizontal directions is very different. In the vertical direction, the refractive indices are very different, resulting in high confinement, but in the horizontal direction the effective indices are much closer.

It must also be remembered that the effective index method is an approximation, and hence there is the question of accuracy. Many authors have considered the accuracy of the method and produced examples of good agreement with more rigorous theory, or relatively poor agreement (see for example [1,2,3]). The general conclusions are that the method becomes inaccurate for complex structures, poorly confined modes, or large index steps; hence care must be exercised using this, or indeed any other, approximate method.

4.4   LARGE SINGLE-MODE RIB WAVEGUIDES

In Chapter 2 we introduced an equation for the approximate number of modes in a planar waveguide (equation 2.29). For a waveguide with dimensions of several microns the number of modes will be large. For example, consider a waveguide of height $h = 5\,\mu m$. Equation 2.29b is reproduced below as equation 4.6:

$$m_{max} = \frac{k_0 n_1 h \cos \theta_c}{\pi} \qquad (4.6)$$

For simplicity assume a symmetrical silicon waveguide, with $n_1 = 3.5$ and $n_2 = 1.5$. Thus, for an operating wavelength of 1.3 μm:

$$\theta_c = \sin^{-1} \frac{1.5}{3.5} = 25.4° \qquad (4.7)$$

Hence:

$$[m_{max}]_{int} = \frac{2 \times 3.5 \times 5 \times 10^{-6} \cos \theta_c}{\lambda_0} = 24 \qquad (4.8)$$

Hence 25 modes are supported (including the $m = 0$ mode).

Alternatively we could consider how thin we need to make the waveguide to support a single mode. Equation 2.33 was the condition for only a single mode to propagate. This equation is reproduced as equation 4.9:

$$\theta_c < \cos^{-1}\left(\frac{\lambda_0}{2n_1 h}\right) \tag{4.9}$$

Rearranging 4.9 in terms of waveguide height h, and substituting for $\cos\theta_c$:

$$h < \left(\frac{\lambda_0/2}{n_1\sqrt{1 - (n_2/n_1)^2}}\right) \tag{4.10}$$

Therefore, for single-mode operation in an SOI waveguide, at a wavelength of 1.3 μm, h needs to be less than approximately 0.2 μm.

It is perhaps a little surprising to discover, then, that rib waveguides are used routinely with silicon layer thicknesses of several microns, and it is routinely claimed that these waveguides are single-mode. Furthermore, this seems directly to contradict the effective index analysis that we have just undertaken, because, although we did not find propagation constants for modes of higher order than the fundamental, clearly we could have done so.

The answer to this apparent paradox is that, if the geometry of the rib waveguide is correctly designed, higher-order modes leak out of the waveguide over a very short distance, leaving only the fundamental mode propagating. This was demonstrated theoretically by Soref et al. [4], in an excellent research paper in 1991. Whilst the entire theory is not recounted here, a summary of their approach is given, together with the crucial design equations.

Firstly the authors defined a rib waveguide in terms of some normalising parameters, as shown in Figure 4.6. The analysis is intended for large rib waveguides that satisfy the condition:

$$2b\sqrt{n_1^2 - n_2^2} \geq 1 \tag{4.11}$$

The waveguide modes were denoted using the convention $HE_{p,q}$ and $EH_{p,q}$ for $p = 0, 1, 2, \ldots$ and $q = 0, 1, 2, \ldots$. The authors then limited their analysis to waveguides in which $0.5 \leq r < 1.0$. This restriction was imposed because, for $r \geq 0.5$, the effective index of 'vertical modes' in the planar region each side of the rib becomes higher than the effective index of all vertical modes in the rib, other than the fundamental.

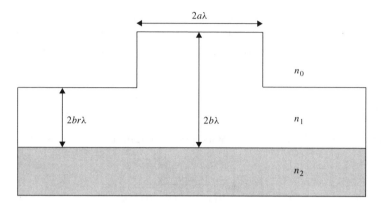

Figure 4.6 Rib waveguide definitions

Thus all modes other than the fundamental vertical mode are cut off. Intuitively this seems reasonable as the second vertical mode in the rib will have an electric field profile with two peaks, the latter overlapping well with the fundamental mode of the planar region adjacent to it, and hence coupling to it. The authors then used an effective index approach to define a parameter related to the aspect ratio of the rib waveguide, a/b. They then found the limiting condition such that the EH_{01} and HE_{01} modes (and hence higher-order modes as well) just failed to be guided. This resulted in a condition for the aspect ratio:

$$\frac{a}{b} \leq 0.3 + \frac{r}{\sqrt{1 - r^2}} \qquad (4.12)$$

Therefore restrictions on the geometry of a rib waveguide were presented which make the guide behave as a single-mode waveguide.

In order to demonstrate their hypothesis the authors simulated the progression of a higher-order mode. Their results are shown in Figure 4.7. Rather conveniently, the waveguide simulated represented a silicon-on-insulator structure, with an upper air cladding. The first figure represents the launch conditions, deliberately off-centre in order to excite higher-order modes. The second diagram shows waveguide modes after travelling $250\,\mu m$, which clearly exhibit more than a single mode. After $2000\,\mu m$ the mode profile is essentially that of only the fundamental mode. Whilst this is not proof of the concept, it is clearly a demonstration that in this case the hypothesis is reasonable, and that the particular higher-order modes present here have leaked away after travelling $2\,mm$.

Since the publication of this work, many authors have demonstrated single-mode waveguides, notably Schmidtchen et al. [5] and

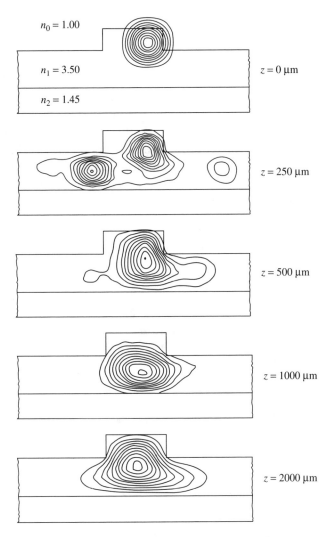

Figure 4.7 Beam propagation simulation of a rib waveguide. Source: R A Soref, J Schmidtchen and K Petermann (1991) 'Large single-mode rib waveguides in GeSi–Si and Si-on-SiO$_2$', *J. Quant. Electron.*, **27**, © 2003 IEEE

Rickman et al. [6], and the notion of large-dimension, single-mode ribs is widely accepted.

However, the precise conditions for single-mode behaviour have been considered by other authors. For example, Pogossian et al. [7] pointed out that a direct effective index approach to the single-mode condition yielded an equation slightly different from that of Soref et al.'s equation (4.12 above). Therefore they took the experimental data of reference [6], and fitted an equation of the form:

$$\frac{a}{b} \leq c + \frac{r}{\sqrt{1 - r^2}} \qquad (4.13)$$

Clearly this is equation 4.12, with the constant 0.3 replaced by a variable c. The direct effective index method suggests that the constant should be zero. The work of Pogossian et al. produced a value of -0.05, much closer to the result by the effective index method (EIM). Their results are shown in Figure 4.8. It can be seen that the agreement is particularly good. Thus the conclusion of this work was that the following equation should be used to ensure single-mode behaviour:

$$\frac{a}{b} \leq \frac{r}{\sqrt{1 - r^2}} \qquad (4.14)$$

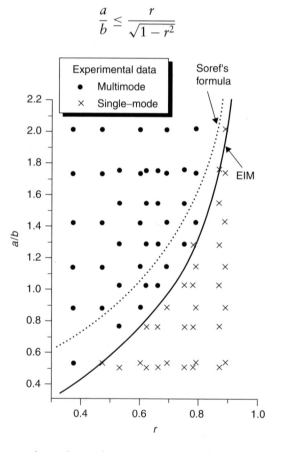

Figure 4.8 The single-mode condition compared to experimental data. Source: S Pogossian, L Vescan and A Vonsovici (1998) 'The single mode condition for semiconductor rib waveguides with large cross-section', *J. Lightwave Technol.*, **16**, © 2003 IEEE

Clearly this is a slightly more restrictive condition than that of equation 4.12. More recently Powell [8] has analysed rib waveguides with angled walls, in a similar manner, producing a similar analytical equation for such waveguides.

The preceding discussion clearly indicates that the modal characteristics of the rib waveguide are intimately related to the waveguide geometry. In particular, for a fixed waveguide height and width, the etch depth can completely change the modal characteristics. As the etch is made deeper, the waveguide may start to support higher-order modes. This becomes particularly important in applications where a single-mode waveguide is required, such as in an interferometer. Consequently the waveguide should be designed such that it is not close to any single-mode/multimode borderline. When the waveguides become smaller, this becomes increasingly difficult. This problem is discussed further in Chapter 7.

4.5   REFRACTIVE INDEX AND LOSS COEFFICIENT IN OPTICAL WAVEGUIDES

So far we have regarded the refractive index as a real quantity, but in general it is complex. Therefore our references to refractive index thus far have, strictly speaking, referred only to the real part of the index. Complex refractive index can be defined as:

$$n' = n_R + jn_I \qquad (4.15)$$

We know from Chapter 2 (equation 2.19) that the propagation constant can be defined as:

$$k = nk_0 \qquad (4.16)$$

We have previously expressed a propagating field as:

$$E = E_0 e^{j(kz-\omega t)} \qquad (4.17)$$

Therefore if we introduce a complex refractive index to equations 4.16 and 4.17, we obtain:

$$E = E_0 e^{j(k_0 n' z - \omega t)} = E_0 e^{jk_0 n_R z} e^{-k_0 n_I z} e^{-j\omega t} \qquad (4.18)$$

Therefore there is a term $\exp(-k_0 n_I z)$. In the same way that the propagation constant in the z direction is redesignated β, this term is often redesignated $\exp(-\frac{1}{2}\alpha)$. The term α is called the *loss coeffi-cient*. The factor of $\frac{1}{2}$ is included in the definition above because, by convention, α is an intensity loss coefficient. Therefore we can write:

$$I = I_0 e^{-\alpha z} \tag{4.19}$$

Clearly equation 4.19 describes how the intensity decays with propaga-tion distance z, through a material. Complex refractive index and loss coefficient will be further discussed later.

Having defined the loss coefficient, it is worth considering what sort of propagation loss we can tolerate for an integrated optical waveguide. Whilst there is no absolute threshold that makes a particular material technology acceptable, the widely accepted benchmark for loss is of the order of 1 dB/cm. This is because an integrated optical circuit is typically a few centimetres in length. Since an optical loss of 3 dB corresponds to a halving of the optical power, it is clear that a loss of much more than 1 dB/cm will rapidly result in a very poor signal-to-noise ratio at the detector. Add to this additional losses due to coupling to or from the optical circuit, or losses within the circuit not associated with propagation loss, and the situation is exacerbated. Losses for SOI waveguides are typically in the range 0.1–0.5 dB/cm.

4.6 CONTRIBUTIONS TO LOSS IN AN OPTICAL WAVEGUIDE

Losses in an optical waveguide originate from three sources: scattering, absorption and radiation. Within each category are further subdivisions of the contribution to the loss, although the relative contribution of each of these effects is dependent upon the waveguide design and the quality of the material in which the waveguide is fabricated. Each of the contributions to loss will be discussed in turn.

4.6.1 Scattering

Scattering in an optical waveguide can be of two types: volume scattering and interface scattering. Volume scattering is caused by imperfections in the bulk waveguide material, such as voids, contaminant atoms, or crystalline defects. Interface scattering is due to roughness at the interface between the core and the claddings of the waveguide. Usu-ally, in a well-established waveguide technology, volume scattering is negligible, because the material has been improved to a sufficient level prior to fabrication of the waveguide. For new or experimental material systems, however, volume scattering should always be considered. Interface scattering may not be negligible, even for a relatively well-developed material system, because losses can be significant even for relatively smooth interfaces.

Whilst it is now well established that silicon waveguides are low-loss propagation media, one might reasonably be concerned that volume scattering could be a contributor to optical loss, since in several of the fabrication techniques used to produce SOI wafers the potential exists for the introduction of defects, notably via ion implantation. It has been shown that the contribution to volume scattering is related to the number of defects, their size with respect to the wavelength of propagation, and the correlation length along the waveguide. In bulk media, Rayleigh scattering is the dominant loss mechanism (see for example [9]), which exhibits a λ^{-4} dependence. However, for confined waves the wavelength dependence is related to the axial correlation length of the defects [10]. For correlation lengths shorter than or of the order of the wavelength, the scattering loss exhibits a λ^{-3} dependence, because the reduction of confinement for longer wavelengths partially counters the λ^{-4} relation. For long correlation lengths compared to the wavelength, radiation losses dominate and a λ^{-1} dependence is observed.

Interface scattering has been studied by a number of authors, who have published a range of expressions for approximating the scattering from the surface or interface of an optical waveguide. However, since many of these models are very complex, an approximate technique will be introduced here, which is attractive owing to its simplicity. The theory of this technique was first introduced by Tien in 1971 [11], and is based upon the specular reflection of power from a surface. This condition holds for long correlation lengths, which is a reasonable assumption in most cases.

If the incident beam has power P_i, the specular reflected power, P_r, from a surface is given by [12]:

$$P_r = P_i \exp\left[-\left(\frac{4\pi\sigma n_1}{\lambda_0}\cos\theta_1\right)^2\right] \qquad (4.20)$$

where σ is the variance of the surface roughness (or r.m.s. roughness), θ_1 is the propagation angle within the waveguide, and n_1 is the refractive index of the core. By considering the total power flow over a given distance, together with the loss at both waveguide interfaces based upon equation 4.20, Tien produced the following expression for the loss coefficient due to interface scattering:

$$\alpha_s = \frac{\cos^3\theta}{2\sin\theta}\left(\frac{4\pi n_1(\sigma_u^2 + \sigma_\ell^2)^{\frac{1}{2}}}{\lambda_0}\right)^2\left(\frac{1}{h + \dfrac{1}{k_{yu}} + \dfrac{1}{k_{y\ell}}}\right) \tag{4.21}$$

where σ_u is the r.m.s. roughness for the upper waveguide interface, σ_ℓ is the r.m.s. roughness for the lower waveguide interface, k_{yu} is the y-directed decay constant in the upper cladding (as defined in Chapter 2), $k_{y\ell}$ is the y-directed decay constant in the lower cladding, and h is the waveguide thickness.

The approximate degree of interface scattering can be evaluated from equation 4.21. This is best demonstrated by an example.

Example

Consider a planar waveguide, as depicted several times in Chapter 2, with the following parameters: $n_1 = 3.5$; $n_2 = 1.5$; $n_3 = 1.0$; $h = 1.0\,\mu m$, and the operating wavelength $\lambda_0 = 1.3\,\mu m$. Let us compare the scattering loss of two different modes of the waveguide, say the TE_0 and the TE_2 modes.

Solution

We can evaluate the propagation angle θ_1 of these two modes as was demonstrated in Chapter 2. The angles are $80.8°$ (TE_0) and $60.7°$ (TE_2). From Chapter 2 we also know that the decay constants in the claddings of the waveguide are given by:

$$k_{yi}^2 = \beta^2 - k_0^2 n_i^2 \tag{4.22}$$

Therefore we can evaluate the decay constant for the upper cladding, k_{yu}, and for the the lower cladding, $k_{y\ell}$, for each mode as in Table 4.1.

If we now let both σ_u and σ_ℓ be 1 nm, we can evaluate the scattering loss for each mode, for each nanometre of r.m.s. roughness at each interface. Therefore, *using equation 4.21*, for the TE_0 mode we have:

$$\alpha_s = 0.04\,cm^{-1} \tag{4.23}$$

Table 4.1

TE$_0$	TE$_2$
$k_{yu} = \sqrt{\beta^2 - k_0^2 n_3^2} = 15.98\,\mu\text{m}^{-1}$	$k_{yu} = \sqrt{\beta^2 - k_0^2 n_3^2} = 13.94\,\mu\text{m}^{-1}$
$k_{y\ell} = \sqrt{\beta^2 - k_0^2 n_2^2} = 15.04\,\mu\text{m}^{-1}$	$k_{y\ell} = \sqrt{\beta^2 - k_0^2 n_2^2} = 12.85\,\mu\text{m}^{-1}$

Expressed in decibels, this is equivalent to a loss of 0.17 dB/cm. For the TE$_2$ mode we have:

$$\alpha_s = 1.33\,\text{cm}^{-1} \tag{4.24}$$

Expressed in decibels, this is equivalent to a loss of 5.78 dB/cm.

It is clear from this example that the higher-order modes will suffer more loss due to interface scattering than the fundamental mode. In this example the TE$_2$ mode experiences more than 30 times more loss than the fundamental mode, resulting from interface scattering. This is due both to the differences in optical confinement, and to more reflections per unit length in the direction of propagation for the higher-order modes.

4.6.2 Absorption

The two main potential sources of absorption loss for semiconductor waveguides are band edge absorption (or interband absorption), and free carrier absorption. Interband absorption occurs when photons with energy greater than the band gap are absorbed to excite electrons from the valence band to the conduction band. Therefore to avoid interband absorption, a wavelength must be used that is longer than the absorption edge wavelength of the waveguide material. Silicon is an excellent example material to demonstrate this point. The band edge wavelength of silicon is approximately 1.1 μm, above which silicon is used as a waveguide material. For wavelengths shorter that 1.1 μm, silicon absorbs very strongly, and it is one of the most common materials used for photodetectors in the visible wavelength range, and at very short infrared wavelengths.

The band edge absorption of a material does not mark an abrupt transition from strong absorbence to transparence, so care must be taken when selecting a wavelength for a given technology. For example, the absorption of pure silicon at a wavelength of $\lambda = 1.15\,\mu\text{m}$ results in an attenuation of 2.83 dB/cm, whereas moving to a wavelength of $\lambda = 1.52\,\mu\text{m}$ reduces the loss to 0.004 dB/cm [13]. Therefore, semiconductor waveguides should suffer negligible band edge absorption, provided a suitable wavelength of operation is chosen.

Free carrier absorption, however, may be significant in semiconductor waveguides. The concentration of free carriers will affect both the real and imaginary refractive indices. For devices fabricated in silicon, modulation of the free carrier density is used to deliberately modulate refractive index, an effect that will be discussed in more detail later in this text when optical modulators are considered. Changes in absorption in semiconductors can be described by the well-known Drude–Lorenz equation [14]:

$$\Delta\alpha = \frac{e^3\lambda_0^2}{4\pi^2 c^3 \varepsilon_0 n}\left(\frac{N_e}{\mu_e (m_{ce}^*)^2} + \frac{N_h}{\mu_h (m_{ch}^*)^2}\right) \qquad (4.25)$$

where e is the electronic charge; c is the velocity of light in vacuum; μ_e is the electron mobility; μ_h is the hole mobility; m_{ce}^* is the effective mass of electrons; m_{ch}^* is the effective mass of holes; N_e is the free electron concentration; N_h is the free hole concentration; ε_0 is the permittivity of free space; and λ_0 is the free space wavelength.

Some of the parameters in equation 4.25 are interdependent, so care must be taken when evaluating the effect of free carrier absorption. However, in order to demonstrate the significance of free carrier absorption let us consider the additional absorption introduced by free carriers. Soref and Lorenzo [15] evaluated equation 4.25 for values of N in the range 10^{18} to 10^{20} cm^{-3}, and presented the data graphically, as shown in Figure 4.9. It can be seen that an injected hole and electron concentration of 10^{18} cm^{-3} introduces a total additional loss of approximately 2.5 cm^{-1}. This corresponds to a loss of 10.86 dB/cm, indicating the dramatic effect doping of the semiconductor can have on the loss in a waveguide.

4.6.3 Radiation

Radiation loss from a straight optical waveguide should ideally be negligible. This type of loss implies leakage from the waveguide into the surrounding media, typically the upper or lower cladding, or for a rib waveguide, into the planar region adjacent to the guide. If the waveguide is well designed this loss will not normally be significant, although unwanted perturbations in the waveguide due to, for example, a slightly damaged fabrication mask may cause scattering of light from one mode to another. The second mode may in turn result in some

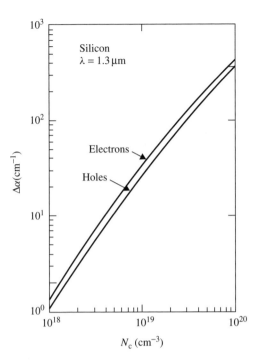

Figure 4.9 Additional loss of silicon due to free carriers. Source: R A Soref and J P Lorenzo (1986) 'All-silicon active and passive guided wave components for $\lambda = 1.3$ and $1.6\,\mu m$', *IEEE J. Quant. Electron.*, **22**, © 2003 IEEE

radiative loss if that mode is leaky. Another situation that may result in some radiative loss is curvature of the waveguide, as this will change the angle of incidence at the waveguide wall, which may in turn result in some radiative loss, as discussed in section 6.3 in Chapter 6.

For a multilayer waveguide structure such as the SOI rib waveguide depicted in Figure 4.2, the possibility of radiative loss exists if the lower waveguide cladding is finite. In the case of SOI, the buried oxide layer must be sufficiently thick to prevent optical modes from penetrating the oxide layer and coupling to the silicon substrate. Clearly the required thickness will vary from mode to mode, as we have seen that each mode penetrates the cladding to a different depth. Furthermore, the penetration depth also varies with the waveguide dimensions with respect to the wavelength of operation. Many authors have analysed this problem for different material systems. In the case of SOI waveguides, with a surface silicon layer of several microns, the buried oxide thickness needs to be at least $0.4\,\mu m$ for operation in the wavelength range $1.3\text{--}1.6\,\mu m$, to prevent significant loss (e.g. [16]). As the waveguide dimensions reduce,

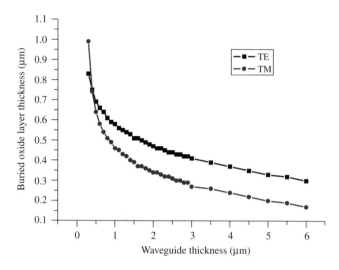

Figure 4.10 Buried oxide layer thickness vs planar SOI waveguide thickness for achieving $\leq 0.001\,\mathrm{dB/cm}$ loss for the fundamental mode at the wavelength of 1550 nm

however, the mode tails will extend proportionately further into the cladding, and it will be necessary to utilise a thicker buried oxide. This is important because there is a trend in silicon photonics to move to smaller waveguide structures for space and efficiency advantages. Consequently it is important to ensure that these advantages are not compromised by making the waveguides more lossy. Hence it is important to ensure that a sufficiently thick buried oxide is utilised.

Figure 4.10 shows how the required buried oxide thickness of a planar SOI waveguide structure varies with a varying waveguide thickness, to maintain a loss to the substrate of less than 0.001 dB/cm of the fundamental mode. Clearly as the waveguide thickness reduces, the required thickness of the buried oxide increases. This is because the mode is becoming less well confined with reducing waveguide dimensions. It is interesting to note the polarisation dependence even at large waveguide thicknesses. This is not usually an issue because the oxide is simply made thick enough to fully confine both polarisations. Note that when the waveguide is reduced to approximately 1 μm in thickness, a buried oxide approaching 1 μm is also required.

4.7 COUPLING TO THE OPTICAL CIRCUIT

Coupling of light to an integrated optical circuit is conceptually trivial, but in practice is a nontrivial problem. This is because of the small size of optical waveguides, typically a few microns at most in any cross-sectional dimension. A variety of techniques exist for performing the coupling task, the most common being end-fire coupling, butt coupling, prism coupling, and grating coupling. End-fire and butt coupling are very similar, involving simply shining light onto the end of the waveguide. The distinction that is usually made between these two methods is that butt coupling involves simply 'butting' the two devices or waveguides up to one another, such that the mode field of the 'transmitting' device falls onto the endface of the second device; whereas end-fire coupling incorporates a lens to focus the input beam onto the endface of the 'receiving' device. Therefore light is introduced into the end of the waveguide, and can potentially excite all modes of the waveguide. Prism coupling and grating coupling, however, are distinctly different approaches, because they introduce an input beam through the surface of a waveguide, at a specific angle. This enables phase matching to a particular propagation constant within the waveguide, thereby enabling excitation of a specific mode. The principles of these four coupling techniques are shown schematically in Figure 4.11.

For the purposes of semiconductor waveguide evaluation, prism coupling is not particularly useful. The conditions of coupling are such that the material from which the prism is made should have a higher refractive index than the waveguide; this seriously restricts the possibilities, particularly for silicon which has a high refractive index of 3.5. There

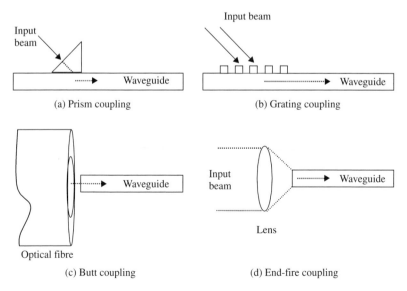

Figure 4.11   Four techniques for coupling light to optical waveguides

are materials available, however, such as germanium which could be used, although other limitations mean that the technique is still inferior to alternative techniques. These limitations are that prism coupling can damage the surface of the waveguide, it is not suitable if a surface cladding is to be used, it is best suited to planar waveguides, and it is certainly not suited to material systems that utilise rib waveguides such as the silicon technology. The remaining three techniques can be useful for silicon-based technology, and will be discussed in turn.

4.7.1 Grating Couplers

Grating couplers provide a means of coupling to individual modes, and they are useful for coupling to waveguide layers of a wide range of thickness. Because the input beam must be introduced at a specific angle, grating couplers are not sufficiently robust for commercial devices, but they are a valuable development tool.

In order to couple light into a waveguide mode, as depicted in Figures 4.11a or b, it is necessary for the components of the phase velocities in the direction of propagation (z direction) to be the same. This is referred to as the *phase-match condition*, and in this case it means that the propagation constants in the z direction must be the same. Consider first a beam (or ray) incident upon the surface of the waveguide at an angle θ_a, as shown in Figure 4.12. The ray will propagate in medium n_3 with a propagation constant $k_0 n_3$, in the direction of propagation. Therefore the z-directed propagation constant in medium n_3 will be:

$$k_z = k_0 n_3 \sin \theta_a \qquad (4.26)$$

Therefore the phase-match condition will be:

$$\beta = k_z = k_0 n_3 \sin \theta_a \qquad (4.27)$$

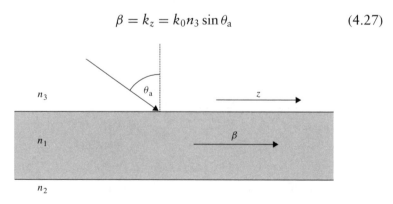

Figure 4.12   Light incident upon the surface of a waveguide

where β is the waveguide propagation constant. However, we know from Chapter 2 (equation 2.39) that $\beta \geq k_0 n_3$. Therefore the condition of equation 4.27 can never be met, since $\sin \theta_a$ will be less than unity. This is why a prism or a grating is required to couple light into the waveguide, because both can satisfy the phase-match condition if correctly designed.

A grating is a periodic structure. If it is to be used as an input or output coupler, it is usual to fabricate the grating on the waveguide surface. The periodic nature of the grating causes a periodic modulation of the effective index of the waveguide. For an optical mode with propagation constant β_W when the grating is not present, the modulation results in a series of possible propagation constants, β_p, given by:

$$\beta_p = \beta_W + \frac{2p\pi}{\Lambda} \tag{4.28}$$

where Λ is the period of the grating, and $p = \pm1, \pm2, \pm3$, etc.

These modes are equivalent to the various diffraction orders of a diffraction grating. Clearly the propagation constants corresponding to positive values of p cannot exist in the waveguide because the propagation constant β_p will still be less than $k_0 n_3$. Therefore only the negative values of p can result in a phase match. It is usual to fabricate the grating such that only the value $p = -1$ results in a phase match with a waveguide mode. Therefore the waveguide propagation constant becomes:

$$\beta_p = \beta_W - \frac{2\pi}{\Lambda} \tag{4.29}$$

Therefore the phase-match condition becomes:

$$\beta_W - \frac{2\pi}{\Lambda} = k_0 n_3 \sin \theta_a \tag{4.30}$$

Writing β_W in terms of the effective index N, equation 4.30 becomes:

$$k_0 N - \frac{2\pi}{\Lambda} = k_0 n_3 \sin \theta_a \tag{4.31}$$

and on substituting for k_0 we obtain:

$$\Lambda = \frac{\lambda}{N - n_3 \sin \theta_a} \tag{4.32a}$$

Since the medium with refractive index n_3 is usually air ($n_3 = 1$), 4.32a becomes:

$$\Lambda = \frac{\lambda}{N - \sin\theta_a} \qquad (4.32b)$$

Equation 4.32b can therefore be used to determine the waveguide grating period for a desired input angle in air for coupling to the mode with propagation constant β_W.

Surprisingly little work has been carried out concerning grating couplers in silicon, mainly owing to the fabrication difficulties. Because the refractive index of silicon is large, the required period of gratings in silicon for input/output coupling is of the order of 400 nm. The highest coupling efficiencies from gratings in silicon have been reported by Ang et al., who achieved an output coupling efficiency of approximately 70 % for rectangular gratings [17], and 84 % for gratings with a nonsymmetrical profile [18]. One of their devices is shown in Figure 4.13.

As a final comment on grating-based devices, it is worth noting that whilst a deep etch perturbs the refractive index of a given waveguide more than does a shallow grating, hence enabling the coupling mechanism, it may also add more loss due to increased scattering. Scattering results from the etch because of the increased abruptness of the refractive

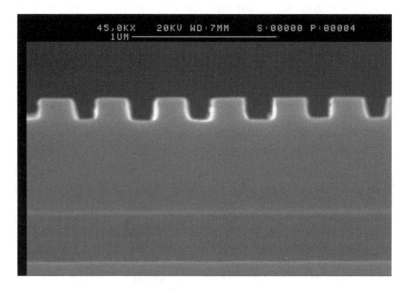

Figure 4.13 Waveguide coupler fabricated in the silicon layer of an SOI waveguide. Source: T W Ang, G T Reed, A Vonsovici et al. 'Grating couplers using silicon on insulator', *Proc. SPIE*, **3620**, (1999). Reproduced by permission of SPIE

index discontinuity between each section of the grating. The fact that silicon has a very large refractive index enhances this effect. Hence it is important to carefully design any grating coupler, ensuring that the most efficient device is used.

4.7.2   Butt Coupling and End-fire Coupling

The parameters that affect the efficiency of butt coupling and end-fire coupling are similar, so these techniques will be discussed together. When a light beam is incident upon the endface of an optical waveguide, the efficiency with which the light is coupled into the waveguide is a function of (i) how well the fields of the excitation and the waveguide modes match; (ii) the degree of reflection from the waveguide facet; (iii) the quality of the waveguide endface; (iv) and the spatial misalignment of the excitation and waveguide fields. There can also be a numerical aperture mismatch in which the input angles of the optical waveguide are not well matched to the range of excitation angles, but this latter term is neglected here.

Overlap of Excitation and Waveguide Fields

Matching of the excitation fields is usually evaluated by carrying out the overlap integral between the excitation field and the waveguide field. In order to evaluate this effect fully, all modes of the waveguide should be included in the waveguide field. However, in practice, many applications require only single-mode operation, and consequently the overlap integral is evaluated between the input field and the fundamental mode of the waveguide. The overlap integral, Γ, of two fields E and ε is given by:

$$\Gamma = \frac{\int_{-\infty}^{\infty} dy \int_{-\infty}^{\infty} E\varepsilon dx}{\left[\int_{-\infty}^{\infty} dy \int_{-\infty}^{\infty} E^2 dx \cdot \int_{-\infty}^{\infty} dy \int_{-\infty}^{\infty} \varepsilon^2 dx\right]^{\frac{1}{2}}} \tag{4.33}$$

where the denominator is simply a normalising factor. The factor Γ lies between 0 and 1, and therefore represents the range between no coupling and total coupling due to field overlap.

A simple way to gain a reasonable idea of the overlap between fields in the excitation and waveguide mode is to use the gaussian approximation introduced in Chapter 3. If the input beam is a gas laser, it is likely to have a gaussian profile which will be well matched to the fundamental mode of a symmetrical waveguide. Similarly a semiconductor laser can be well matched to the fundamental mode of a waveguide, if the device geometries are similar. Even a single-mode fibre will have a near-gaussian profile, as discussed in Chapter 3. Therefore, it is useful to consider an example of the overlap between two gaussian functions. Let the two fields E and ε be described by the following expressions:

$$E = \exp\left[-\left(\frac{x^2}{\omega_x^2} + \frac{y^2}{\omega_y^2}\right)\right] \qquad (4.34)$$

This represents a waveguide field with $1/e$ widths in the x and y directions of $2\omega_x$ and $2\omega_y$ respectively. If we assume a circularly symmetrical input beam, then ε is given by:

$$\varepsilon = \exp\left[-\frac{(x^2 + y^2)}{\omega_0^2}\right] \qquad (4.35)$$

Using the mathematical identity for a definite integral [19]:

$$\int_0^\infty \exp[-r^2 x^2]\, dx = \frac{\sqrt{\pi}}{2r} \qquad (4.36)$$

equation 4.33 reduces to:

$$\Gamma = \frac{\dfrac{2}{\omega_0}\left[\dfrac{1}{\omega_x \omega_y}\right]^{\frac{1}{2}}}{\left[\dfrac{1}{\omega_x^2} + \dfrac{1}{\omega_0^2}\right]^{\frac{1}{2}}\left[\dfrac{1}{\omega_y^2} + \dfrac{1}{\omega_0^2}\right]^{\frac{1}{2}}} \qquad (4.37)$$

Since equation 4.37 is the coupling efficiency of the field profiles, the power coupling efficiency is given by:

$$\Gamma^2 = \frac{\dfrac{4}{\omega_0^2}\left[\dfrac{1}{\omega_x \omega_y}\right]}{\left[\dfrac{1}{\omega_x^2} + \dfrac{1}{\omega_0^2}\right]\left[\dfrac{1}{\omega_y^2} + \dfrac{1}{\omega_0^2}\right]} \qquad (4.38)$$

Table 4.2

ω_0	ω_x	ω_y	Γ	Γ^2	Loss due to Γ^2 (dB)
$5\,\mu m$	$5\,\mu m$	$5\,\mu m$	1.0	1.0	0
$10\,\mu m$	$10\,\mu m$	$5\,\mu m$	0.894	0.8	0.97
$20\,\mu m$	$16\,\mu m$	$3\,\mu m$	0.535	0.286	5.4
$20\,\mu m$	$1\,\mu m$	$1\,\mu m$	0.1	0.01	20
$5\,\mu m$	$5\,\mu m$	$3\,\mu m$	0.939	0.882	0.55
$5\,\mu m$	$4.8\,\mu m$	$4.9\,\mu m$	0.999	0.999	0.004

We can now evaluate equations 4.37 and 4.38 for different values of ω_x, ω_y and ω_0. A range of values are shown in Table 4.2, as verification of the validity of equation 4.38. Obviously the symmetry of the input beam described means that values of ω_x and ω_y can be interchanged to yield the same value for the overlap integral. However, the same approach can be adopted with a different aspect ratio for the input beam, to represent a noncircularly symmetric device such as a semiconductor laser, or another waveguide.

Reflection from the Waveguide Facet

Reflection from the waveguide endface is determined by the refractive indices of the media involved in coupling from one medium to another, and is described by the Fresnel equations introduced in Chapter 2. From equation 2.5 the reflection coefficient r_{TE} for TE polarisation was:

$$r_{TE} = \frac{n_1 \cos \theta_1 - n_2 \cos \theta_2}{n_1 \cos \theta_1 + n_2 \cos \theta_2} \qquad (4.39)$$

Similarly from equation 2.6, for TM polarisation, the reflection coefficient r_{TM} was:

$$r_{TM} = \frac{n_2 \cos \theta_1 - n_1 \cos \theta_2}{n_2 \cos \theta_1 + n_1 \cos \theta_2} \qquad (4.40)$$

We recall that the reflection coefficients refer to reflection of fields, and that to determine the reflection of power we use the reflectivity, R (equation 2.12):

$$R = r^2 \qquad (4.41)$$

Using Snell's law (equation 2.1), equation 4.39 reduces to:

$$r_{TE} = \frac{-\sin(\theta_1 - \theta_2)}{\sin(\theta_1 + \theta_2)} \qquad (4.42)$$

Hence the reflectivity is given by:

$$R_{TE} = r_{TE}^2 = \frac{\sin^2(\theta_1 - \theta_2)}{\sin^2(\theta_1 + \theta_2)} \tag{4.43}$$

Similarly R_{TM} can be found to be:

$$R_{TM} = r_{TM}^2 = \frac{\tan^2(\theta_1 - \theta_2)}{\tan^2(\theta_1 + \theta_2)} \tag{4.44}$$

The two functions are plotted in Figure 4.14 for an air/silicon interface (i.e. $n_1 = 1.0, n_2 = 3.5$). Notice that at normal incidence ($\theta_1 = 0$), the reflection of both TE and TM polarisations is the same. Furthermore, end-fire coupling introduces light at near-normal incidence. Consequently the approximation is usually made that the Fresnel reflection at the waveguide facets is that due to normal incidence. In this case, equations 4.43 and 4.44 both reduce to:

$$R = \left| \frac{n_1 - n_2}{n_1 + n_2} \right|^2 \tag{4.45}$$

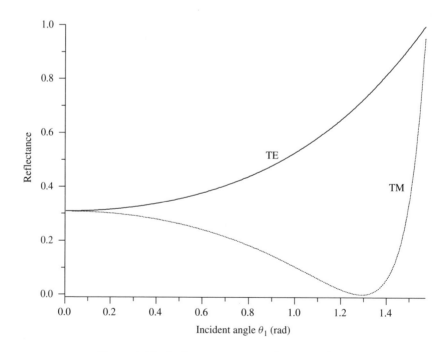

Figure 4.14   Reflection at an air/silicon interface

For a silicon/air interface this reflection is approximately 31 %, which introduces an additional loss of 1.6 dB. A loss of 1.6 dB for each facet of the waveguide is considerable, and is reduced in commercial devices by the use of antireflection coatings.

An antireflection coating is an additional coating between two media which is used to reduce or remove the reflection at the interface. For the case of the facet of an optical waveguide, the facet would be coated in the antireflection coating. The coating thickness is such that the waves reflected from the front of the coating and from the facet are in antiphase (i.e. the thickness is equal to $\lambda/4$), so that they will to some extent cancel, reducing or eliminating the reflection. It can be shown that, for normal incidence, the net reflectivity is given by [20]:

$$R = \left| \frac{n_1 n_2 - n_{ar}^2}{n_1 n_2 + n_{ar}^2} \right|^2 \qquad (4.46)$$

where n_{ar} is the refractive index of the antireflection coating. It can be seen from equation 4.46 that R will be zero if:

$$n_1 n_2 = n_{ar}^2 \qquad (4.47)$$

For a silicon/air interface this means that n_{ar} needs to be approximately 1.87. Silicon nitride (Si_3N_4) has a refractive index of approximately 2.05 and is sometimes used as an antireflection coating. Alternatively silicon oxynitride (SiO_xN_y) can be used to form layers with a variety of refractive indices ranging from 1.46 (SiO_2) to 2.05 (Si_3N_4), by changing the relative oxygen and nitrogen concentrations.

The Quality of the Waveguide Endface

The quality of the waveguide endface is very dependent upon the preparation technique. Three main options are available for endface preparation of semiconductor waveguides: cleaving, polishing and etching.

Cleaving is carried out by mechanically introducing a small crack at the edge of a sample, and subsequently applying pressure to the sample so that is cracks along a primary crystal plane. This technique is difficult to master, and is limited to use in the research laboratory, since the nature of the technique renders it unsuitable for commercial application. The results can be very good, but tend to be variable in SOI because the introduction of a buried oxide layer means the surface and substrate layers of silicon are not directly connected.

Polishing is probably the most common method of preparing a waveguide facet. The sample endface is polished by lapping with abrasive materials with sequentially decreasing grit sizes. Usually a prescribed 'recipe' is followed, which can result in an excellent surface finish, although 'rounding' of the endface is a common failing.

Endfaces can also be prepared by chemical or dry etching. The detail of how this is achieved is beyond the scope of this text, but suffice to say it is a technique that can be developed to a sufficiently high level for commercial application.

Clearly the aim of any surface preparation technique is to produce a sufficiently smooth facet that optical scattering is minimised. In the present context that means that no features approaching the dimensions of the optical waveguide should be present. For the purposes of attenuation measurements, the assumption is usually made that the polishing technique has worked perfectly (or to such a level that scattering is negligible), and consequently no allowance is made for endface scattering loss.

Spatial Misalignment of the Excitation and Waveguide Fields

Alignment of integrated optical components is of critical importance owing to the very small dimensions of the devices involved. To demonstrate this, let us reconsider the overlap of the exciting and the waveguide fields in equation 4.33, reproduced below as 4.48:

$$\Gamma = \frac{\int_{-\infty}^{\infty} dy \int_{-\infty}^{\infty} E\varepsilon dx}{\left[\int_{-\infty}^{\infty} dy \int_{-\infty}^{\infty} E^2 dx \cdot \int_{-\infty}^{\infty} dy \int_{-\infty}^{\infty} \varepsilon^2 dx\right]^{\frac{1}{2}}} \qquad (4.48)$$

where

$$E = \exp\left[-\left(\frac{x^2}{\omega_x^2} + \frac{y^2}{\omega_y^2}\right)\right] \quad \text{and} \quad \varepsilon = \exp\left[-\frac{(x^2 + y^2)}{\omega_0^2}\right]$$

If we now introduce a term A into either E or ε, to represent offset of one field from the other, we can investigate the additional losses small misalignments can introduce. The offset factor can be introduced to either equation since the offset is of one field with respect to the other.

For the sake of argument we will introduce the term into the field ε, so:

$$\varepsilon = \exp\left[-\frac{(x^2 + (y - A)^2)}{\omega_0^2}\right] \tag{4.49}$$

In this case the term A represents an offset in the y direction. This expression can be rewritten as:

$$\varepsilon = \exp\left[-\frac{(x^2 + (y^2 + A^2 - 2Ay))}{\omega_0^2}\right] \tag{4.50}$$

Therefore the overlap integral equation 4.48 can be manipulated into the form:

$$\Gamma^I = \exp\left[-\frac{A^2}{\omega_y^2 + \omega_0^2}\right] \cdot \frac{\int_{-\infty}^{\infty} dy \int_{-\infty}^{\infty} E\varepsilon dx}{\left[\int_{-\infty}^{\infty} dy \int_{-\infty}^{\infty} E^2 dx \cdot \int_{-\infty}^{\infty} dy \int_{-\infty}^{\infty} \varepsilon^2 dx\right]^{\frac{1}{2}}} \tag{4.51}$$

$$= \exp\left[-\frac{A^2}{\omega_y^2 + \omega_0^2}\right]\Gamma \tag{4.52}$$

Therefore we can evaluate an offset weighting parameter ψ

$$\psi = \exp\left[-\frac{A^2}{\omega_y^2 + \omega_0^2}\right] \tag{4.53}$$

which 'weights' the overlap integral due to the offset. The square of this term ψ^2 is also evaluated as it represents the additional loss when considering power transmission (coupling).

The terms are evaluated in Figure 4.15 for a range of offsets A, when $\omega_y = \omega_0 = 5\,\mu\text{m}$. It can be seen that an offset of only $2\,\mu\text{m}$ corresponds to a value of 'ψ^2' of approximately 0.85, which represents an additional loss in the coupling process of about $0.7\,\text{dB}$. Clearly this loss is due to an offset in the y direction only and would be compounded by an additional offset in the x direction.

4.7.3 Robust Coupling to Waveguides for Commercial Applications

The techniques discussed in section 4.7.2 are mainly laboratory techniques, although butt coupling can be used commercially if the fibre

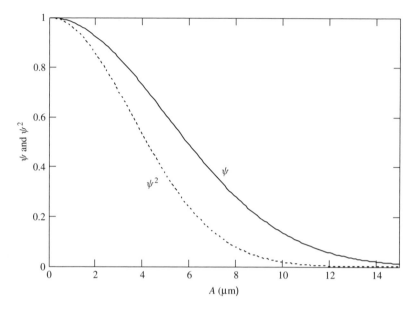

Figure 4.15   Effect of an offset A on the electric field overlap and the power coupling efficiency

and waveguide have similar mode sizes. However, all coupling techniques become more difficult as the waveguide dimensions are reduced. For silicon waveguides above approximately $2\,\mu m$ in cross-sectional dimensions, some form of three-dimensional taper can be used but for smaller waveguides the loss is prohibitively large. The three-dimensional tapered waveguide transition can, in theory, offer a monolithically integrated means by which efficient coupling can be achieved. The three-dimensional taper is a gradual transition from a large cross-sectional waveguide area to a smaller one. Even for waveguides of several microns in cross-section, a taper can be advantageous, because it relaxes the alignment tolerances between the input fibre and the waveguide. The aim is to produce a taper that reduces the waveguide dimensions in a smooth, lossless transition. The angle of the taper is typically very small to achieve the smooth transition, and it must also be produced with very low surface roughness. Lateral tapers are relatively straightforward to fabricate because this is essentially just an etching process from the top of the silicon wafer, but to produce a vertical taper requires a differential etch rate along the length of the taper. A schematic of a three-dimensional taper is shown in Figure 4.16.

Such vertical tapers have been used in experimental situations, but Bookham Technology have successfully commercialised an alternative

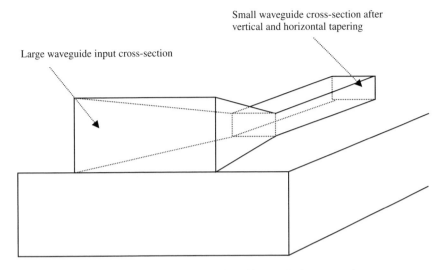

Small waveguide cross-section after vertical and horizontal tapering

Large waveguide input cross-section

Figure 4.16 Schematic of vertical and horizontal waveguide tapers

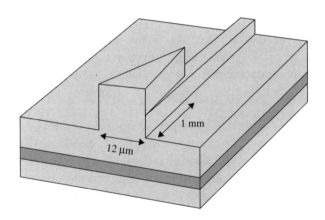

1 mm

12 μm

Figure 4.17 NVT taper to aid waveguide coupling, from Bookham Technology. Reproduced from I Day, I Evans, A Knights et al. (2003) 'Tapered silicon waveguides for low-insertion-loss highly efficient high-speed electronic variable attenuators', in *IEEE OFC 2003*, 24–27 March by permission of Optical Society of America

tapered structure, which is shown in Figure 4.17 [21]. They call their structure an NVT mode-matching taper, which matches the fibre mode to the input mode of the taper structure. The large waveguide is then tapered to a point, transferring power to the underlying rib waveguide. They also report very little polarisation dependence, and an insertion loss of less than 0.5 dB/facet, which is shown in Figure 4.18. The latter is not unexpected owing to the large cross-sectional area of

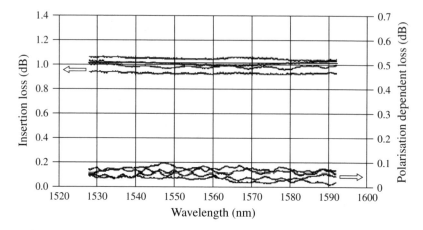

Figure 4.18 Fibre-to-fibre performance of two of Bookham Technology's NVT taper, connecting the two fibres to a variable optical attenuator. Reproduced from I Day, I Evans, A Knights et al. (2003) 'Tapered silicon waveguides for low-insertion-loss highly efficient high-speed electronic variable attenuators', in *IEEE OFC 2003*, 24–27 March by permission of Optical Society of America

the input of the device. Details are not given in published data of the exact size of the waveguide that the taper reduces to, but the commercial devices to which it is coupled, together with misalignment tolerances quoted, suggest waveguides of several microns in cross-section.

The Bookham Technology taper is an elegant device, because it removes the need to fabricate a vertical taper, and hence also removes the difficulty of achieving a smooth vertical taper transition. However, it does require some means of forming the large vertical structure above the rib waveguide. The published data indicate that this is locally epitaxially grown onto the SOI waveguide, which significantly complicates the fabrication process.

It is hard to imagine a commercial process that does not require some form of taper to ease the difficulty of coupling to the optical waveguides. The Bookham Technology NVT device is one solution, and a more conventional vertical taper is another, and both have their own complexities.

However, the trend to smaller waveguide dimensions will increase the coupling difficulties of these and other devices. For example, when considering the vertical Si taper to couple light from a fibre to a small waveguide, it is achievable in theory, but it is very difficult to produce a practical vertical taper of good quality which has low radiation loss. The technique seems to work well for waveguides larger than about

2 μm in cross-sectional dimensions, but smaller dimension prove much more difficult.

Alternatively some form of grating-based device can be used to transfer power from air to small waveguides. Surface grating couplers, discussed in section 4.7.1, offer one means of coupling to small waveguides. However, such an approach is not sufficiently robust for commercial applications, and other approaches must be used. Chapter 6 discusses devices used for coupling to small waveguides further, including three-dimensional tapers, and a grating-based coupler.

4.7.4 Measurement of Propagation Loss in Integrated Optical Waveguides

Having considered the contributions to both loss and coupling efficiency, it is worth briefly considering common methods of measuring the loss of optical waveguides. Several techniques are available for waveguide loss measurement, and the method implemented will depend upon the information required from the experiment. For example, if the loss associated with particular modes of a waveguide is required, then clearly a coupling technique that is mode selective is required, such as prism coupling or grating coupling. Alternatively, if the total loss of the waveguide is required, end-fire coupling or butt coupling are attractive owing to the simplicity with which either technique can be implemented.

Insertion Loss and Propagation Loss

When an optical waveguide is measured, the objective of the measurement must be made clear. There is often confusion between insertion loss and propagation loss. The insertion loss of a waveguide or device is the total loss associated with introducing that element into a system, and therefore includes both the inherent loss of the waveguide itself and the coupling losses associated with exciting the device. This measurement is typically associated with system design or specification, since the total loss of introducing the device clearly affects the system performance.

Alternatively, the propagation loss is the loss associated with propagation in the waveguide or device, excluding coupling losses. To optimise the performance of a given device during development or research, it is usually the propagation loss that is of interest, since the contributions to the loss due to waveguide design or material properties are associated with the propagation loss. It is sometimes necessary to derive one of these parameters from the other.

There are three main experimental techniques associated with waveguide measurement: (i) the cut-back method; (ii) the Fabry–Perot resonance method; and (iii) scattered light measurement. Each method will be described in turn.

The Cut-back Method

The cut-back method is conceptually the simplest method of measuring an optical waveguide, and is usually associated with either end-fire coupling or butt coupling. A waveguide of length L_1 is excited by one of the coupling methods mentioned, and the output power from the waveguide, I_1, and the input power to the waveguide, I_0, are recorded. The waveguide is then shortened to another length, L_2, and the measurement repeated to determine I_2, whilst the input power I_0 is kept constant. The propagation loss of the length of waveguide $(L_1 - L_2)$ is therefore related to the difference in the output power from each measurement. Using equation 4.19 we can express this as:

$$\frac{I_1}{I_2} = \exp[-\alpha(L_1 - L_2)] \tag{4.54}$$

so that

$$\alpha = \left(\frac{1}{L_1 - L_2}\right) \ln(I_2/I_1) \tag{4.55}$$

Note that this expression has been produced from only two data points, and has assumed that the input coupling, the condition of the waveguide endfaces, and the input power all remain constant. The accuracy of the technique can be improved by taking multiple measurements and plotting a graph of the optical loss against waveguide length, as shown in Figure 4.19. Note that the data presented are now insertion loss for each device length, not propagation loss.

Whilst Figure 4.19 does not depict real data, it is usual to fit a straight line to the experimental data, the degree of scatter representing the uncertainty associated with the measurements. Notice that the dotted line representing a straight-line fit to this data cuts the y-axis above the origin. Clearly this suggests a loss for zero propagation length. This is because the data are expressed as insertion loss rather than propagation loss, and hence the loss corresponding to zero propagation length is the total coupling loss to the waveguide.

As a slight variation to this technique, input coupling can be arranged to be via other techniques such as prism coupling, because the coupling

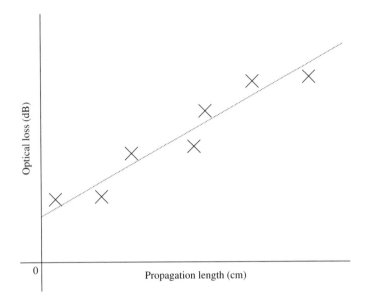

Figure 4.19   Scatter of optical loss measurements with waveguide length

loss itself is removed from the data expressed as propagation loss. The important consideration then becomes which coupling technique is the most reproducible for a variety of waveguide lengths.

A second variation of the cut-back method, although inherently less accurate, is to carry out an insertion loss measurement for a single waveguide length, and calculate the coupling loss by evaluating the reflection from each waveguide endface, together with the mode field mismatch, as described in section 4.7.2. Whilst this technique is usually a little less accurate than the cut-back method, it has the enormous advantage of being nondestructive.

The Fabry–Perot Resonance Method

An optical waveguide with polished endfaces (facets) is similar in structure to the cavity of a laser. Light propagates along the waveguide, and may be reflected at either facet, by an amount determined by the refractive index of the waveguide material and the external media (usually air), as described by equations 4.43 to 4.45. However, any coating on the waveguide facet may change this reflectivity.

Therefore the waveguide structure may be regarded as a resonant cavity, with the waves undergoing multiple reflections as they pass along the waveguide and back. Such a cavity is called a *Fabry–Perot cavity*.

The optical intensity transmitted through such a cavity, I_t, is related to the incident light intensity, I_0, by the well-known equation:

$$\frac{I_t}{I_0} = \frac{(1-R)^2 e^{-\alpha L}}{(1-Re^{-\alpha L})^2 + 4Re^{-\alpha L}\sin^2(\phi/2)} \tag{4.56}$$

where R is the facet reflectivity, L is the waveguide length, α is the loss coefficient, and ϕ is the phase difference between successive waves in the cavity. Equation 4.56 has a maximum value when $\phi = 0$ (or multiples of 2π), and a minimum value when $\phi = \pi$; that is:

$$\frac{I_{max}}{I_0} = \frac{(1-R)^2 e^{-\alpha L}}{(1-Re^{-\alpha L})^2} \tag{4.57a}$$

$$\frac{I_{min}}{I_0} = \frac{(1-R)^2 e^{-\alpha L}}{(1-Re^{-\alpha L})^2 + 4Re^{-\alpha L}} = \frac{(1-R)^2 e^{-\alpha L}}{(1+Re^{-\alpha L})^2} \tag{4.57b}$$

Therefore we can evaluate the ratio of the maximum intensity to minimum intensity as:

$$\zeta = \frac{I_{max}}{I_{min}} = \frac{(1+Re^{-\alpha L})^2}{(1-Re^{-\alpha L})^2} \tag{4.58}$$

We can rearrange equation 4.58 as:

$$\alpha = -\frac{1}{L}\ln\left(\frac{1}{R}\frac{\sqrt{\zeta}-1}{\sqrt{\zeta}+1}\right) \tag{4.59}$$

Therefore if we know the reflectivity R, and if we can measure the ratio of the maximum intensity to minimum intensity ζ, the loss coefficient can be evaluated.

The transfer function of the Fabry–Perot cavity is described by equation 4.56, and is periodic when the phase ϕ passes through multiples of 2π. Therefore if we can experimentally sweep the Fabry–Perot cavity through a few cycles of 2π, ζ can be measured. Such cycling can be achieved by varying the temperature of the waveguide and hence the length, or alternatively by varying the wavelength of the light source.

Equation 4.56 is plotted for three different reflectivities in Figure 4.20. The values of R are 0.1, 0.31 and 0.5, and $\alpha L = 0.023$ (corresponding to a loss of approximately 0.1 dB/cm). We can see from Figure 4.20 that, for low reflectivity ($R = 0.1$), the response has a near sinusoidal shape. As the reflectivity increases, the response has sharper peaks ($R = 0.31$,

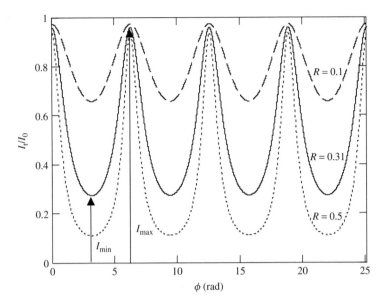

Figure 4.20   Plot of the Fabry–Perot transfer function for three different mirror reflectivities

0.5). The value of 0.31 is of particular significance as it corresponds to an air/silicon interface, so the response is similar to what we may expect from a real silicon waveguide. Also marked on Figure 4.20 are the maximum and minimum values of the transfer function, I_{max} and I_{min}, as used in equations 4.56–4.59. The reader should confirm that by inserting the values of I_{max} and I_{min} into equation 4.59, a value of $\alpha L = 0.023$ can be obtained.

Notice from equation 4.59 that the coupling efficiency does not affect the calculation of loss coefficient, although it is implicit that the coupling efficiency remains constant. For this reason, varying the temperature of the waveguide can be problematic if it causes any variation in coupling to the waveguide, so sweeping the wavelength of the light source is preferable, although more complex.

This method can be particularly useful for measuring low losses, under certain conditions. The accuracy of the technique has been compared with other methods by Tittelbach et al. [22], who demonstrated that the technique is accurate if the reflectivity is known to better than 0.01, and is relatively high, of the order of 0.6. Under these circumstances, an attenuation–length product in the range 0.05–0.1 can be measured well. For silicon waveguides, αL is typically in the range 0.02–0.8. At the lower end of this range the Fabry–Perot method is reliable, but towards the middle and upper end the cut-back method gives smaller errors.

A Fabry–Perot scan can also give additional information. A Fabry–Perot cavity with relatively low reflectivity at the facets (approximately 31 % for an air/silicon interface) results in a transfer function similar to the solid curve in Figure 4.20, as discussed above. However, if the waveguide is multimode, then additional interference causes distortion of this response and hence gives information about whether the waveguide is single-mode, slightly multimode, or highly multimode. Of course the latter two situations require some interpretation of the result, and are not quantitative concerning the number of modes within the waveguide. For example, Figure 4.21 shows a relatively 'clean' Fabry–Perot scan, indicating a single-mode device. Figure 4.22 shows a multimode device scan. Both scans were provided by the Intel Corporation. It is clear that Figure 4.21 is similar to the solid curve of Figure 4.20. Furthermore, the trace in Figure 4.22 is significantly distorted compared to that in Figure 4.21, and is not the expected transfer function of a typical Fabry–Perot cavity for a single-mode device. Consequently it is reasonable to conclude that the waveguide is multimode, although without additional modelling little quantitative information is provided of additional modes. Nevertheless, it is often convenient to confirm the existence of a single-mode response or otherwise.

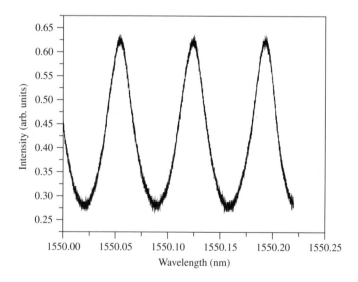

Figure 4.21 Fabry–Perot scan of a single-mode waveguide. Reproduced from Data provided by Dr R Nicolaescu, Intel Corporation by permission of Intel Corporation

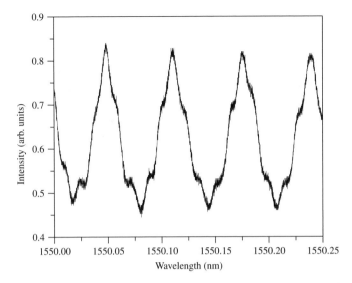

Figure 4.22 Fabry–Perot scan of a multimode waveguide. Reproduced from Data provided by Dr R Nicolaescu, Intel Corporation by permission of Intel Corporation

Scattered Light Measurement

The measurement of scattered light from the surface of a waveguide can also be used to determine the loss. The assumption underlying this method is that the amount of light scattered is proportional to the propagating light. Therefore, if scattered light is measured as a function of waveguide length, the rate of decay of scattered light will mimic the rate of decay of light in the waveguide. Optical fibres can be used to collect scattered light from the surface of a waveguide, and can be scanned along the surface. Alternatively an image of the entire surface can be made, and the decay of scattered light determined accordingly. However, it is clear that light is scattered significantly only if the loss of the waveguide is high, and relatively high power is propagating in the waveguide. In many situations, neither is desirable, and hence this method tends to be used for initial studies of waveguide materials, when losses are high, but is discarded in favour of other techniques when the waveguide has been optimised.

4.8 OPTICAL MODULATION MECHANISMS IN SILICON

One of the requirements of an integrated optical technology, particularly one related to communications, is the ability to perform optical modulation. This implies a change in the optical field due to some applied signal, typically (although not exclusively) an electrical signal. The change in the optical field is usually derived from a change in refractive index of the material involved, with the applied field, although other parametric changes are possible. It is now widely accepted that the most efficient means of implementing optical modulation in silicon via an electrical signal is to use carrier injection or depletion. This will be discussed later. Firstly, however, let us consider other electrical modulation techniques to discover why they are not useful in silicon, even though they are used in other integrated optical technologies. The primary candidates for electrically derived modulation are electric field effects.

4.8.1   Electric Field Effects

The application of an electric field to a material can result in a change to the real and imaginary refractive indices. A change in real refractive index, Δn, with an applied electric field is known as *electrorefraction*, and a change in the imaginary part of refractive index, $\Delta \alpha$, with applied electric field is known as *electroabsorption*. The primary electric field effects that are useful in semiconductor materials to cause either electroabsorption or electrorefraction are the Pockels effect, the Kerr effect, and the Franz–Keldysh effect. Soref and Bennett [24] have examined electric field effects in silicon, and their results will be cited here to demonstrate the relative efficiency of several electric field effects. However, let us first briefly discuss the three electric field effects mentioned above.

The Pockels Effect

The Pockels effect, also known as the linear electro-optic effect, causes a change in real refractive index, Δn, which is proportional to the applied field, E. Therefore, if the applied field is uniform, the change in refractive index will be proportional to applied voltage, V, for a fixed modulator geometry. The Pockels effect in general produces a change in refractive index that is dependent upon the direction of the applied electric field with respect to the crystal axes. Therefore the effect is also usually polarisation dependent, although simplifications occur owing to the symmetry of the crystal structure of a given material, in any given direction. The geometry of the silicon crystal structure is such that the Pockels effect disappears completely, so it is not an option for optical modulation in silicon. However, more generally it is usual to align the applied electric field with one of the principal axes of the crystal, to utilise the largest electro-optic coefficient for the material concerned. For example, for the material lithium niobate (LiNbO$_3$), the maximum refractive index change occurs if the so-called 'r_{33}' coefficient is used. This results in a refractive index change in LiNbO$_3$ given by:

$$\Delta n = -r_{33}n_{33}\frac{E_3}{2} \qquad (4.60)$$

where n_{33} is the refractive index in the direction of the applied electric field, and E_3 is the applied electric field. The subscript 3 merely indicates which of the three principal axes of the material is aligned with the applied field. The value of r_{33} is 30.8×10^{-12} m/V.

The Kerr Effect

The Kerr effect is a second-order electric field effect in which the change in real refractive index, Δn, is proportional to the square of the applied electric field. This effect is present in silicon, although it is relatively weak. The change may be expressed as:

$$\Delta n = s_{33}n_0\frac{E^2}{2} \qquad (4.61)$$

where s_{33} is the Kerr coefficient, n_0 is the unperturbed refractive index, and E is the applied field. In this case the sign of the refractive index change is not dependent on the direction within the crystal axis.

Soref and Bennett [24] theoretically quantified the refractive index change in silicon at a wavelength of 1.3 μm, as a function of the applied electric field. Their results are shown in Figure 4.23. It can be seen that the change in refractive index, Δn, is predicted to reach 10^{-4} at an applied field of 10^6 V/cm (100 V/μm), which is above the breakdown field in lightly doped silicon. It will be seen later that this is a relatively small effect when compared to the plasma dispersion effect (carrier injection).

The Franz–Keldysh Effect

Unlike the Pockels effect and the Kerr effect, the Franz–Keldysh effect gives rise to both electrorefraction and electroabsorption, although primarily the latter. The effect is due to distortion of the energy bands of the

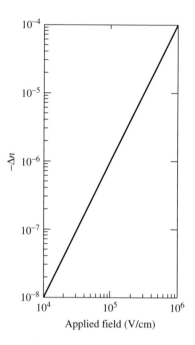

Figure 4.23   The Kerr effect in silicon as a function of applied electric field, at 300 K and $\lambda = 1.3\,\mu$m. Source: R A Soref and B R Bennett (1987) 'Electrooptical effects in silicon', *IEEE J. Quant. Electron.*, **23**, © 2003 IEEE

semiconductor upon application of an electric field. In effect, this shifts the bandgap energy, resulting in a change in the absorption properties of the crystal, particularly at wavelengths close to the bandgap, and hence a change in the complex refractive index. Soref and Bennett [24] also quantified the changes in refractive index due to the Franz–Keldysh effect. Whilst not all their data were presented graphically, they plotted the change in refractive index as a function of the applied electric field at a wavelength of $1.07\,\mu$m, the wavelength at which the effect is greatest. They also plotted data at $1.09\,\mu$m for comparison. It should be noted, however, that the effect diminishes significantly at the telecommunications wavelengths of $1.31\,\mu$m and $1.55\,\mu$m. The data are shown in Figure 4.24.

It can be seen that the refractive index change reaches 10^{-4} at an applied field of 2×10^5 V/cm (20 V/μm). Whilst this figure is better than that evaluated for the Kerr effect, it must be remembered that this is not evaluated at the same wavelength. The Franz–Keldysh effect will fall very significantly at $1.3\,\mu$m, the wavelength at which the results above for the Kerr effect are quoted.

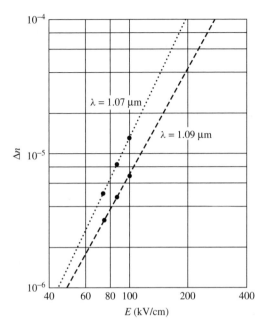

Figure 4.24 The Franz–Keldysh effect in silicon at 300 K and for two values of λ. Source: R A Soref and B R Bennett (1987) 'Electrooptical effects in silicon', *IEEE J. Quant. Electron.*, **23**, © 2003 IEEE

4.8.2 Carrier Injection or Depletion

We observed in section 4.6 that the concentration of free charges in silicon contributes to the loss via absorption. We also observed in section 4.5 that the imaginary part of the refractive index is determined by the absorption (or loss) coefficient. Therefore it is clear that changing the concentration of free charges can change the refractive index of the material. We saw from section 4.6 that the Drude–Lorenz equation relating the concentration of electrons and holes to the absorption (equation 4.25) was:

$$\Delta\alpha = \frac{e^3\lambda_0^2}{4\pi^2c^3\varepsilon_0 n}\left(\frac{N_e}{\mu_e(m_{ce}^*)^2} + \frac{N_h}{\mu_h(m_{ch}^*)^2}\right) \tag{4.62}$$

The corresponding equation relating the carrier concentrations, N, to change in refractive index, Δn, is [14]:

$$\Delta n = \frac{-e^2\lambda_0^2}{8\pi^2c^2\varepsilon_0 n}\left(\frac{N_e}{m_{ce}^*} + \frac{N_h}{m_{ch}^*}\right) \tag{4.63}$$

Soref and Bennett [24] studied results in the scientific literature to evaluate the change in refractive index, Δn, to experimentally produced absorption curves for a wide range of electron and hole densities, over a wide range of wavelengths. In particular they focused on the communications wavelengths of 1.3 μm and 1.55 μm. Interestingly their results were in good agreement with the classical Drude–Lorenz model, only for electrons. For holes they noted a $(\Delta N)^{0.8}$ dependence. They also quantified the changes that they had identified from the literature, for both changes in refractive index and in absorption. They produced the following extremely useful expressions, which are now used almost universally to evaluate changes due to injection or depletion of carriers in silicon:

At $\lambda_0 = 1.55$ μm:

$$\Delta n = \Delta n_e + \Delta n_h = -[8.8 \times 10^{-22} \Delta N_e + 8.5 \times 10^{-18} (\Delta N_h)^{0.8}] \quad (4.64)$$

$$\Delta \alpha = \Delta \alpha_e + \Delta \alpha_h = 8.5 \times 10^{-18} \Delta N_e + 6.0 \times 10^{-18} \Delta N_h \quad (4.65)$$

At $\lambda_0 = 1.3$ μm:

$$\Delta n = \Delta n_e + \Delta n_h = -[6.2 \times 10^{-22} \Delta N_e + 6.0 \times 10^{-18} (\Delta N_h)^{0.8}] \quad (4.66)$$

$$\Delta \alpha = \Delta \alpha_e + \Delta \alpha_h = 6.0 \times 10^{-18} \Delta N_e + 4.0 \times 10^{-18} \Delta N_h \quad (4.67)$$

where: Δn_e = change in refractive index resulting from change in free electron carrier concentrations; Δn_h = change in refractive index resulting from change in free hole carrier concentrations; $\Delta \alpha_e$ = change in absorption resulting from change in free electron carrier concentrations; $\Delta \alpha_h$ = change in absorption resulting from change in free hole carrier concentrations.

We will see later that a carrier injection level of the order of 5×10^{17} is readily achievable for both electrons and holes. Therefore, if we evaluate the change in refractive index, Δn, at a wavelength of 1.3 μm (equation 4.66), for comparison with the field effects evaluated earlier, we obtain:

$$\Delta n = -[6.2 \times 10^{-22} (5 \times 10^{17}) + 6.0 \times 10^{-18} (5 \times 10^{17})^{0.8}]$$

$$= -1.17 \times 10^{-3} \quad (4.68)$$

This is more than an order of magnitude larger than the changes due to the electric field effects described earlier. Furthermore, by appropriate high doping of contacts to the silicon layer, it is reasonable to expect higher injection levels, and therefore higher refractive index changes.

4.8.3   The Thermo-optic Effect

In addition to the electric field effects and injection of free carriers into silicon, one other modulation technique has proved viable for optical modulation devices in silicon. It is the thermo-optic effect, in which the refractive index of silicon is varied by applying heat to the material. The thermo-optic coefficient in silicon is [25]:

$$\frac{\mathrm{d}n}{\mathrm{d}T} = 1.86 \times 10^{-4}/\mathrm{K} \tag{4.69}$$

Therefore if the waveguide material can be raised in temperature by approximately $6\,^\circ$C in a controllable manner, a refractive index change of 1.1×10^{-3} results. There are of course issues about controlling the temperature rise to the locality of the waveguide, and of efficiency of the mechanism used to deliver the thermal energy. However, experimental results [25] suggest that a 500-μm device length can deliver a phase shift of π radians for an applied power of $10\,$mW, if the waveguide is thermally isolated from the substrate. This corresponds to a thermal change of approximately $7\,^\circ$C, and hence a refractive index change of approximately 1.3×10^{-3} over the length of the device.

It is worth noting that the refractive index change is positive with applied thermal energy, whereas the injection of free carriers results in a decrease in refractive index. Therefore the two effects could compete in a poorly designed modulator.

4.9   OTHER ADVANTAGES AND DISADVANTAGES OF SILICON PHOTONICS

Some of the advantages, and perhaps some of the disadvantages, of silicon will be obvious from the foregoing discussions of both material and device characteristics. However, it is useful to list the advantages and disadvantages of the technology, in the light of what we have learnt in this text so far (see Table 4.3).

Whilst most of the advantages and disadvantages are self explanatory, let us consider in a little more detail the advantage labelled 'viii' in the table. The term 'micromaching' refers to the fact that silicon is

Table 4.3

Advantages of silicon photonics	Disadvantages of silicon photonics
i. Stable, well-understood material ii. Stable native oxide available for cladding/electrical isolation iii. Relatively low-cost substrates iv. Optically transparent at important wavelengths of 1.3 μm and 1.55 μm v. Well-characterised processing vi. Highly confining optical technology vii. High refractive index means short devices viii. Micromachining means V-grooves and an effective hybrid technology are possible ix. Semiconductor material offers the potential of optical and electronic integration x. High thermal conductivity means high-power devices or high packing density may be tolerated xi. Carrier injection means optical modulation is possible xii. Thermo-optic effect means a second possibility for optical modulation exists	i. No Pockels effect ii. indirect bandgap means native optical sources are not possible iii. High refractive index means inherently short devices are difficult to fabricate (e.g. gratings) iv. Modulation mechanisms tend to be relatively slow v. Thermal effects can be problematic for some optical circuits

sufficiently versatile that very accurate etching of the material is possible such that a variety of interesting shapes can be produced in a silicon substrate, on the micron scale. Aside from the optoelectronic applications of micromachined silicon, a large range of micromechanisms have been fabricated including valves, springs, nozzles, and an array of sensors.

The crucial stage of micromachining is the etching stage. Two types of etch are available for micromachining. *Isotropic etchants* attack the silicon crystal at the same rate in all directions, and therefore tend to make rounded shapes. *Anisotropic etchants* work at different rates in different directions in the crystal lattice. Consequently, well-defined shapes can be formed by masking parts of the silicon to the etch. Typical isotropic etchants are mixtures of hydrofluoric, nitric and acetic acids. Typical anisotropic etches are hot alkaline solutions such as aqueous potassium hydroxide (KOH), aqueous sodium hydroxide (NaOH), and EDP (a mixture of ethylenediamine, pyrocatechol and water). Alternatively, dry etching techniques can be used, such as reactive ion etching, which also tend to be anisotropic etches as they are typically related to the angle at which the etching beam or plasma impinges on the silicon.

Anisotropic etchants are the most useful for silicon micromachining because they allow specific shapes to be etched. Since the etch rate varies with crystal orientation, it is important to know the crystallographic axes of the silicon being used. Whilst it is beyond the scope of this text to deal with detailed crystallography, it is worth briefly considering the silicon crystal structure.

Each silicon atom bonds with four neighbouring atoms, such that each atom lies at the centre of a tetrahedron defined by the four neighbours. A useful diagram from Angell et al. [26] is reproduced as Figure 4.25. It is perhaps easier to visualise the structure when considering more

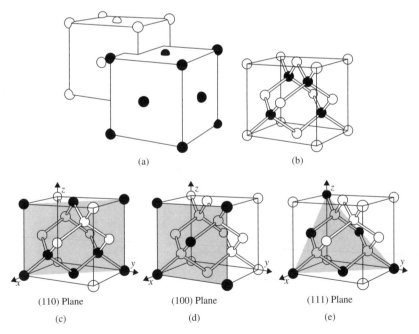

(a)   (b)

(110) Plane   (100) Plane   (111) Plane
(c)   (d)   (e)

Figure 4.25   Silicon crystal structure. Reproduced from J B Angell, S C Terry and P B Barth (1983) 'Silicon micromechanical devices', *Sci. Amer.* **248**, 36–47 by permission of Gabor Kiss, New York

than five atoms, since the larger structure can be viewed as a series of interlocking cubes, each being known as a unit cube (Figure 4.25b). Each unit cube has an atom at each corner, and at the centre of each face, and interlocks with four neighbouring cubes. This is referred to as a face-centred cubic structure. The basic unit of measure is the length of the edge of one cube. The direction in the crystal is described by three coordinates known as the Miller indices (see for example [27]). This allows definition of specific crystal planes, or directions within the crystal. Consider Figure 4.25d. The shaded plane intersects the x-axis at a point one unit measurement from the origin. Using the notation (x, y, z) this is denoted the (100) plane, and the associated direction normal to this plane is denoted [100]. Similarly Figure 4.25c shows the (110) plane, and Figure 4.25e the (111) plane. It is also worth noting that a family of planes, denoted {100}, are commonly defined, which means those planes that have a single intercept along one of the principal axes of the crystal (i.e. (100), (010) and (001)). The associated family of directions normal to this family of planes is then denoted ⟨100⟩.

Whilst the different crystallographic planes etch at different rates, the precise etch rates depend on the composition of the etchant, the temperature, and the silicon face exposed to the etchant. However, in general the {111} planes tend to etch slowly, the {110} planes rapidly, and the {100} planes at an intermediate rate [26]. Another diagram from Angell et al. [26] is reproduced below which demonstrates some basic etching strategies. Obviously the shape of an etched feature further depends on the mask used. Figure 4.26a shows a rounded hole etched by an isotropic etchant. Figure 4.26b shows a square opening along the ⟨110⟩ direction in a mask, on a ⟨100⟩ wafer, yielding a pyramidal pit with {111} sides. Figure 4.26c shows a similar but larger opening

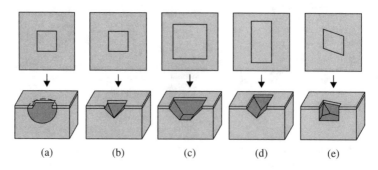

(a)   (b)   (c)   (d)   (e)

Figure 4.26 Etching of silicon via isotropic and anisotropic etches. Reproduced from J B Angell, S C Terry and P B Barth (1983) 'Silicon micromechanical devices', *Sci. Amer.* **248**, 36–47 by permission of Gabor Kiss, New York

results in a flat-bottomed pit, because the etch has been stopped before intersection of the {111} planes has been reached. Figure 4.26d shows how a rectangular opening in the mask on a similarly oriented wafer can result in a V-groove (often used to align waveguides and optical fibres). Finally Figure 4.26e shows a ⟨110⟩ wafer with an opening resulting in a more complex shape.

An example schematic of a complex shape etched into a silicon wafer is shown in Figure 4.27. This device was developed for gas sensing, and includes 1.5-m capillary etches into the silicon surface. The channel was sealed by bonding glass to the surface. The operation of this device is not discussed here, but it is included merely as a demonstration of silicon micromachining.

Clearly the example shown in Figure 4.27 is not associated with silicon photonics. However, we saw another example of micromachining earlier in Figure 4.13, when discussing grating couplers. The grating coupler

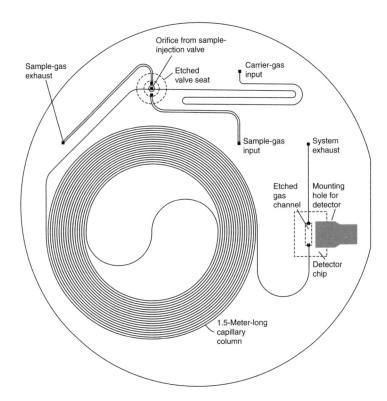

Figure 4.27 Example schematic of a complex shape etched into a silicon wafer. Reproduced from J B Angell, S C Terry and P B Barth (1983) 'Silicon micromechanical devices', *Sci. Amer.* **248**, 36–47 by permission of Gabor Kiss, New York

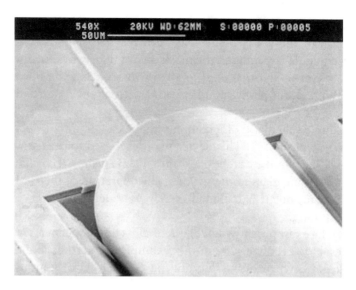

Figure 4.28   Optical fibre in a V-groove. Reproduced by permission of Bookham Technology PLC

was produced by using a silicon dioxide mask, and reactive ion etching, rather than chemical etching. The impressive features of that grating are that the period of the grating is only 400 nm, and the sides of the etched pits are near vertical.

Another common feature in silicon-based photonics is the etched V-groove, into which optical fibres are fixed in order to achieve good alignment with optical waveguides. An example is shown in Figure 4.28; this was produced by Bookham Technology PLC. Numerous other examples exist of micromaching, a subject that has become a large research topic as well as a commercial industry.

REFERENCES

1. H Kogelnik (1988) 'Theory of optical waveguides', in T Tamir (ed.), *Guided Wave Optoelectronics*, Springer-Verlag, Berlin.
2. T M Benson, R J Bozeat and P C Kendall (1992) 'Rigorous effective index method for semiconductor waveguides', *Proc. IEE*, **139**, 67–70.
3. A Kumar, D F Clark and B Culshaw (1988) 'Explanation of errors inherent in the effective index method for analysing rectangular core waveguides', *Opt. Lett.*, **13**, 1129–1131.
4. R A Soref, J Schmidtchen and K Petermann (1991) 'Large single-mode rib waveguides in GeSi–Si and Si-on-SiO$_2$', *J. Quant. Electron.*, **27**, 1971–1974.

5. J Schmidtchen, A Splett, B Schüppert and K Petermann (1991) 'Low-loss single-mode optical waveguide with large cross-section in SOI', *Electron. Lett.*, **27**, 1486–1487.

6. A G Rickman, G T Reed and F Namavar (1994) 'Silicon on insulator optical rib waveguide loss and mode characteristics', *J. Lightwave Technol.*, **12**, 1771–1776.

7. S Pogossian, L Vescan and A Vonsovici (1998) 'The single mode condition for semiconductor rib waveguides with large cross-section', *J. Lightwave Technol.*, **16**, 1851–1853.

8. O Powell (2002) 'Single mode condition for silicon rib waveguides,' *J. Lightwave Technol.*, **20**, 1851–1855.

9. J M Senior (1992) *Optical Fiber Communications: Principles and Practice*, 2nd edn, Prentice-Hall, London.

10. J Nayyer, Y Suematsu and H Tokiwa (1975) 'Mode coupling and radiation loss of clad type optical waveguides due to the index inhomogeneities of the core material', *Opt. Quant. Electron.*, **7**, 481–492.

11. P K Tien (1971) 'Light waves in thin films and integrated optics', *Appl. Opt.*, **10**, 2395–2413.

12. R G Hunsperger (1991) *Integrated Optics: Theory and Technology*, 3rd edn, Springer-Verlag, Berlin.

13. Emis Data Review Series no. 4 (1988) *Properties of Silicon*, INSPEC (IEE), London, pp. 72–79.

14. T S Moss, G J Burell and B Ellis (1973) *Semiconductor Opto-electronics*, Butterworth, London.

15. R A Soref and J P Lorenzo (1986) 'All-silicon active and passive guided wave components for $\lambda = 1.3$ and $1.6\,\mu m$', *IEEE J. Quant. Electron.*, **QE-22**, 873–879.

16. G T Reed, A G Rickman, B L Weiss et al. (1992) 'Optical characteristics of planar waveguides in SIMOX structures', *Proc. Mater. Res. Soc.*, **244**, 387–393.

17. T W Ang, G T Reed, A Vonsovici et al. (1999) 'Grating couplers using silicon on insulator', *Proc. SPIE*, **3620**, 79–86.

18. T W Ang, G T Reed, A Vonsovici et al. (1999) 'Blazed grating coupler in Unibond SOI', *Proc. SPIE*, **3896**, 360–368.

19. H B Dwight (1961) *Tables of Integrals and Other Mathematical Data*, 4th edn, Macmillan, New York.

20. See for example E Hecht (1998) *Optics*, 3rd edn, Addison-Wesley, Reading, MA, ch. 9.

21. I Day, I Evans, A Knights et al. (2003) 'Tapered silicon waveguides for low-insertion-loss highly efficient high-speed electronic variable attenuators', in *IEEE OFC 2003*, 24–27 March.

22. G Tittelbach, B Richter and W Karthe (1993) 'Comparison of three transmission methods for integrated optical waveguide propagation loss measurement', *Pure Appl. Opt.*, **2**, 683–706.

23. Data provided by Dr R Nicolaescu, Intel Corporation, San José, California.

24. R A Soref and B R Bennett (1987) 'Electrooptical effects in silicon', *IEEE J. Quant. Electron.*, **QE-23**, 123–129.

25. S A Clark, B Culshaw, E J C Dawney and I E Day (2000) 'Thermo-optic phase modulators in SIMOX material', *Proc. SPIE*, **3936**, 16–24.

26. J B Angell, S C Terry and P B Barth (1983) 'Silicon micromechanical devices', *Sci. Amer.* **248**, 36–47.

27. See for example S M Sze (1981) *Physics of Semiconductor Devices*, 2nd edn, John Wiley & Sons, New York, ch. 1.

5

Fabrication of Silicon Waveguide Devices

The most attractive aspects of fabricating optical devices from silicon are the low primary cost of the material, the mature and well-characterised processing techniques that are underpinned by decades of research, development and manufacturing in the microelectronics industry, and the potential for straightforward integration with electrical components in the same substrate. In this chapter we turn to the practical realities of making silicon photonic devices, from the specifications for starting material through to the manufacture of devices with optical and electrical integration. We do not attempt to offer a comprehensive review of silicon processing which can be obtained from a number of excellent books (for example [1]), but rather to introduce the reader to the most important fabrication processes involved in silicon photonics.

5.1 SILICON-ON-INSULATOR (SOI)

The vast majority of silicon photonic devices have been fabricated using the SOI platform. SOI is a generic term used to describe structures which consist of a thin layer of crystalline silicon on an insulating layer. The most common SOI structure found in microelectronics is silicon-on-silicon dioxide (SiO_2), in which a uniform layer of SiO_2 is sandwiched between a thick (hundreds of microns) silicon substrate and a thin surface layer of crystalline silicon (Figure 5.1). The thicknesses of both the crystalline silicon and the buried silicon dioxide layer are of the

Silicon Photonics: An Introduction   Graham T. Reed and Andrew P. Knights
© 2004 John Wiley & Sons, Ltd   ISBN: 0-470-87034-6

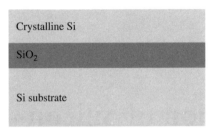

Figure 5.1   Schematic of silicon-on-silicon dioxide

order of a micron, but this value can be varied significantly depending on the fabrication process.

The buried SiO_2 has a refractive index of 1.46, significantly lower than that for the crystalline silicon layer of approximately 3.5. Hence, this type of SOI forms a classic waveguide structure. The waveguide can be made symmetric by thermal oxidation of the surface (see 5.3 below). In the following three subsections we describe the most common methods for the manufacture of SOI substrates which have been used successfully to fabricate silicon waveguides.

5.1.1   Separation by IMplanted OXygen (SIMOX)

The SIMOX process has proved the most popular method for the fabrication of large volumes of SOI material. Although the process is simple in concept, there exists a relatively small process window for the production of device-grade SOI. The key to the fabrication of SOI by the SIMOX method is the implantation of a large number of oxygen ions below the surface of a silicon wafer. Ion implantation is dealt with in more detail in section 5.5.

As we shall see later, the most convenient way to describe the total amount of any ion species implanted into a wafer is by the implanted ion dose. The dose is simply the total number of ions that pass through one square centimetre of the wafer surface and is measured in units of $ions/cm^2$. The total implantation dose required in the SIMOX process is usually $>10^{18}$ cm^{-2}, and under normal room-temperature conditions an unwanted amorphous silicon overlayer would form during the implantation of the oxygen ions. To prevent this, the silicon substrate is maintained at a temperature of approximately $600\,°C$ during implantation. To form an idea of the relative size of the dose required for SIMOX, we can compare it with the dose implanted in a typical doping process for the formation of source and drain contacts in the fabrication of a complimentary metal-oxide semiconductor (CMOS) transistor, usually the highest dose used for electrical doping. To meet the specifications for low electrical resistance, boron (p-type dopant) and arsenic (n-type dopant) would normally be implanted to a dose of no greater than 10^{16} cm^{-2}, fully two orders of magnitude less than in the SIMOX process.

Returning to the description of SIMOX, we note that the oxygen ions are implanted into crystalline silicon at an energy of up to 200 keV. This energy subsequently determines the depth of the SiO$_2$ and hence the thickness of the silicon overlayer. The development of the oxygen profile as a function of depth from the silicon wafer surface is shown schematically in Figure 5.2. The profiles shown are merely indicators of the development of oxygen concentration versus depth as the process proceeds. Experimental measurements of oxygen profiles can be found in [2] and references therein. In common with all implanted species, at low doses ($<10^{16}$ cm^{-2}), the profile can be described by a shape that is similar, but not identical, to that of a gaussian function (Figure 5.2a). As the implantation proceeds further and the dose increases, the peak concentration of oxygen ions (O$^+$), saturates to a concentration of that found in stoichiometric SiO$_2$ (Figure 5.2b). With further implantation the oxygen profile begins to flatten, forming a buried, continuous layer of SiO$_2$.

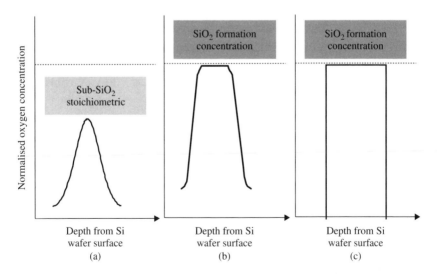

Figure 5.2 Variation of the oxygen profile during the SIMOX process. (a) Low-dose; (b) high-dose (peak is at the stoichiometric limit for SiO$_2$); and (c) after implantation and annealing at 1300 °C for several hours

Following oxygen implantation to a dose of $>10^{18}\,\text{cm}^{-2}$, the silicon wafer is annealed at a temperature of approximately $1300\,^\circ\text{C}$ for several hours. This anneal produces a uniform, buried SiO_2 layer with distinct interfaces with the two adjacent silicon layers. The annealing ensures the silicon overlayer is denuded of implantation-related, primary lattice defects (Figure 5.2c).

The material quality of SIMOX-SOI has been the subject of a great deal of research. Of importance to silicon photonics is the concentration of secondary defects in the silicon overlayer (for example crystal dislocations which could influence the optical propagation properties of a silicon waveguide), and the micro-roughness of the silicon overlayer surface and the overlayer/buried oxide interface. In the 1980s, during the development of a production-worthy SIMOX process, the concentration of dislocations was approximately $10^{10}\,\text{cm}^{-2}$ [3], several orders of magnitude too high for use in the microelectronics industry. This has been reduced to values approaching $10^2-10^3\,\text{cm}^{-2}$, by optimizing fabrication conditions such as implantation substrate temperature and post-implantation annealing conditions. Silicon overlayer surface micro-roughness and overlayer/buried SiO_2 interface micro-roughness are discussed in detail in section 5.4, where we consider their importance for the development of submicron waveguide devices.

The depth and thickness of the buried SiO_2 layer is a function of the implant energy. For a $200\,\text{keV}$ implant the buried SiO_2 thickness is approximately $0.5\,\mu\text{m}$ with a crystalline silicon overlayer of $0.3\,\mu\text{m}$. To make the dimensions of this structure more suitable for the fabrication of large cross-section waveguides, the silicon thickness can be increased by up to several microns via epitaxial growth [4]. Epitaxially grown silicon, used routinely in microelectronics fabrication, has a low concentration of both lattice defects and doping impurities and is dealt with in section 5.1.4.

SIMOX wafers are commercially available from a number of suppliers and can be purchased pre- or post-epitaxial growth. Typical uniformities of buried oxide and silicon overlayer thicknesses across the area of a wafer approach the few percent level.

5.1.2   Bond and Etch-back SOI (BESOI)

Bringing two hydrophilic surfaces (such as SiO_2) into intimate contact can result in the formation of a very strong bond. This phenomenon led to the development of the BESOI process in the early 1970s.

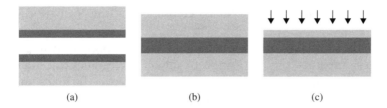

Figure 5.3   The bond and etch-back process to form BESOI: (a) oxidation; (b) bonding; and (c) thinning

The production of BESOI has three steps shown schematically in Figure 5.3: (a) oxidation of the two wafers to be bonded; (b) formation of the chemical bond; (c) thinning (etching) of one of the wafers.

The details of the bonding chemistry are complicated and beyond the scope of this text. In general, the wafers are brought into contact at room temperature, at which point an initial bond is formed. The bond strength is increased to that of bulk material via subsequent thermal processing to temperatures as high as $1100\,^\circ C$.

Wafer thinning can be achieved via a number of different processes. The most common is *chemical mechanical polishing* (CMP), a technique used extensively in microelectronics for wafer planarization. CMP requires that the wafer surface be both weakened and subsequently removed during a single processing step. In general, the silicon surface to be polished is brought into contact with a rotating pad, and simultaneously a chemically reactive slurry containing an abrasive component such as alumina and glycerine weakens and removes surface layers. The process removes the majority of one of the bonded wafers, leaving a thin silicon overlayer on a buried SiO_2 layer, supported by a silicon substrate. The rather crude method of removing hundreds of microns of silicon limits the thicknesses achievable for the silicon overlayer to around 10 microns [3].

An improvement in SOI thickness uniformity (and hence reduced silicon overlayer thickness) can be achieved by the use of an end-stop in the thinning process reducing or even eliminating the need for CMP. This technique is described in detail in [5]. Before bonding of the two oxidised wafers takes place, one of the wafers is doped heavily p-type (i.e. to a concentration $>10^{19}\,cm^{-3}$) via implantation or epitaxial growth. A highly selective etch (i.e. one that will not etch p-doped silicon), sometimes in combination with CMP, is used to remove the majority of the wafer with the p-type doped layer. As the doped region is exposed to the etchant the etching process ceases, leaving a silicon overlayer whose depth and uniformity are determined by the formation of the doped layer.

A modified version of this process has been used previously to produce wafers suitable for the fabrication of silicon waveguides [6]. Following the creation of the heavily doped p-type layer, a further, undoped intrinsic layer is epitaxially grown on the wafer surface. Further, a second nonselective etch process is used to remove the exposed p-type layer following the selective etch. The final wafer structure is then one of an undoped silicon overlayer on the buried SiO_2. As we have seen in Chapter 4, section 4.6.2, silicon waveguides require doping levels $<10^{16} \, cm^{-3}$, so the etch-stop consisting of dopant at a concentration $>10^{19} \, cm^{-3}$ must be removed from the wafer structure if the end-stop technique is to be used.

5.1.3   Wafer Splitting (SmartCut® Process to Produce Unibond® Wafers)

A process which possesses steps from both the SIMOX and BESOI processes is SmartCut® [7], which was developed a decade ago at LETI, France. The process is shown schematically in Figure 5.4.

A thermally oxidised wafer is implanted with hydrogen to a dose of approximately $10^{17} \, cm^{-2}$. The implanted hydrogen ions form a gaussian-like profile. The distance from the wafer surface of the peak of the profile depends on the H^+ ion energy, but is usually between a few hundred nanometres and a few microns. The H^+ ions, and the silicon lattice damage caused by the stopping of the ions, are at their greatest concentration at this depth, and here the silicon lattice bonds are significantly weakened.

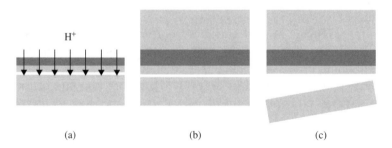

Figure 5.4   The SmartCut® process. (a) Thermally oxidised wafer is implanted with a high dose (approximately $10^{17} \, cm^{-2}$) of hydrogen. (b) A second wafer is bonded to the first as in the BESOI process. (c) Thermal processing splits the implanted wafer at a point consistent with the range of the hydrogen ions

Following implant, the wafer is brought into contact with a second, handling wafer (which may or may not have a thermal SiO_2-covered surface). Upon contact, room-temperature bonding occurs, such as that for the BESOI process. Subsequent thermal processing at 600 °C and 1100 °C first splits the implanted wafer along the peak of the hydrogen implantation profile, then strengthens the bond between the implanted and handling wafer. A fine CMP is employed to reduce roughness at the SOI surface.

Subsequent epitaxial silicon growth can increase the thickness of the silicon overlayer if required. The nonuniformity of the position of the implanted hydrogen profile peak and hence the overlayer thickness are only a few percent. Importantly, although the overlayer receives a high dose implant, the small mass of the hydrogen ions ensures that negligible residual damage remains at the end of the thermal processing. A value for threading dislocations at a concentration $<10^2 \, cm^{-2}$ has been reported [8]. The buried SiO_2 thickness is defined by the thermal oxidation process and hence may be varied from nanometres to several microns.

The flexibility, high quality and efficient use of silicon offered by this process makes it an excellent platform for the development of low-cost silicon photonics [9]. SOI wafers formed via the SmartCut® process are available commercially from SOITEC.

5.1.4 Silicon Epitaxial Growth

For the fabrication of large cross-section (several square microns) silicon waveguides [4], SIMOX and UniBond® wafers do not in general have the thickness of silicon overlayer required. However, it is possible to use the silicon overlayer as a seed layer for the epitaxial growth of silicon to the required thickness. 'Epitaxial' means that the grown layer is an ordered mono-crystal, essential if efficient micro- or opto-electrical devices are to be fabricated. The use of epitaxial silicon as the waveguiding medium has the additional advantage of doping and defect levels below those found in wafers cut from an ingot following bulk growth using the Czochralski (Cz) or float zone (FZ) methods [1].

The most common epitaxial silicon growth technique is *chemical vapour deposition* (CVD). CVD is a process which deposits a solid film on the surface of a silicon wafer by the reaction of a gas mixture at that surface. For silicon deposition, dichlorosilane (SiH_2Cl_2) is often used as the source gas. It is a requirement that the wafer surface be raised in temperature (typically >1000 °C) to provide energy to drive the chemical reaction. Although either a vapour source or a solid source, as with molecular beam epitaxy (MBE), may be used, the desired thickness of the film required in silicon waveguide fabrication dictates that vapour phase epitaxy be dominant. The uniformity and repeatability of the commercially available machines and processes can result in thickness nonuniformities being less than 1 %.

An explanation of the gas transport and reaction kinetics of CVD is provided in [2], while [1] describes in some detail the various processes and equipment available. For the purposes of this text we merely wish to bring to the reader's attention the ability of epitaxial growth to engineer the thickness of the silicon overlayer following the manufacture of the base SOI material.

5.1.5 Deciding on the SOI

Which type of SOI is most suitable for silicon waveguide fabrication? The answer to this question is not straightforward and depends on many factors such as the device application, the manufacturing budget, and the numbers of devices to be produced. All of the material types (SIMOX, BESOI and UniBond®) have been used in the successful manufacture of silicon waveguides. However, there are differences in the SOI structures which should be considered at the commencement of a project.

At the time of writing, SIMOX-SOI has been commercially available for nearly two decades. It is arguably the most researched structure of the three material types, and because it is based on an ion implantation process it is consistently reproducible in respect to quality and dimension. Some concerns exist regarding the properties of the interface between the silicon overlayer and the buried oxide (i.e. micro-roughness) if SIMOX-SOI is to be used in the production of submicron waveguides, but this has yet to be quantified. Certainly there are no such concerns related to waveguides of large cross-section. For the production of silicon overlayers greater than a few hundred nanometres, it is necessary to combine the SIMOX technique with silicon epitaxy.

The advantage of BESOI compared to SIMOX is a silicon overlayer which has not been subjected to high-dose oxygen implantation, and the flexibility available in defining both the overlayer and the buried SiO_2 dimensions. Both may be fabricated with thicknesses of several microns. However, because of the inherent destructiveness of the thinning process, two start wafers are required to produce one SOI wafer using the BESOI process. In high-volume manufacture this can be a drawback.

Without the use of an etch-stop, BESOI wafers have significantly greater nonuniformities than those produced via SIMOX and SmartCut®.

SOI produced via SmartCut® (Unibond® wafers) combines the repeatability of an ion implantation process with the flexibility of BESOI in allowing the buried oxide thickness and silicon overlayer to be varied within wide dimensional limits (up to a few microns). This means that an epitaxial growth may not be necessary to form the waveguide structure if the desired maximum height of the silicon waveguide is less than approximately 2 μm. As with the SIMOX process, the silicon overlayer will have been subjected to a high-dose implantation, although the use of H^+ as opposed to O^+ ions reduces the risk of residual defects in the SOI structure. Unlike the BESOI process, the wafer from which the silicon overlayer is split can be recycled, so only one start wafer is required to produce one SOI wafer.

5.2   FABRICATION OF SURFACE ETCHED FEATURES

Section 5.1 dealt with the fabrication of SOI material and in particular with various approaches to forming silicon-on-SiO_2. Such structures can be used to guide light; but without definition of features which provide lateral confinement, these slab waveguides have few applications in silicon photonics. The physics of a silicon rib waveguide, and the restrictions on its dimensions to provide single-mode operation, were addressed in Chapter 4. In this section we look at the fabrication steps in forming the rib and other guiding structures in the silicon overlayer.

The simplest possible surface structure one can envisage is shown schematically in Figure 5.5a. Extended into the plane of the paper, this is representative of a silicon rib waveguide with vertical walls. Figure 5.5b is a micrograph of an actual SOI rib waveguide. In this case, a deposited SiO_2 layer provides an upper cladding, while just discernable is the buried oxide layer providing the lower cladding.

5.2.1   Photolithography

The width of a rib waveguide (such as that in Figure 5.5) is primarily determined by photolithographical definition, so control of the photo process is one of the most important challenges in silicon photonic fabrication. Furthermore, it is not sufficient to ensure local control, but to maintain both inter- and intra-wafer dimensions within a tight tolerance over many months or years of operation. Although such

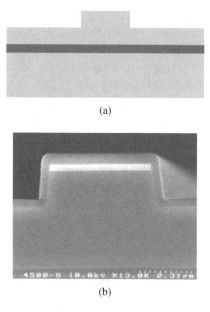

(a)

(b)

Figure 5.5 (a) Schematic of a silicon rib waveguide. (b) Electron micrograph of a silicon rib waveguide. Reproduced by permission of Intel Corporation

manufacturing discipline is required for all fabrication processes, a large number of process variables make photolithography one of the most demanding disciplines in silicon device fabrication.

Current photolithography capabilities are described in the *International Technology Semiconductor Roadmap* [10]. The microelectronics industry has been working with deep-submicron technologies for many years. Attainable minimum feature sizes are now well below 100 nm with critical dimension (CD) control at the 10 nm level. This kind of control is in excess of that required to form the most basic of silicon photonic structures such as the large-cross-section, single-mode silicon rib where minimum feature size is several microns and, by careful choice of the target dimensions, CD control can be relaxed to several hundred nanometres [11]. However, the silicon photonics engineer may require the very best control available when fabricating more exotic devices or waveguides with dimensions which are submicron.

Photolithography transfers a mask-defined pattern to the surface of a wafer. The pattern is printed on the wafer using a photosensitive polymer referred to as *photoresist*. Many recipes and process variations have been developed to ensure good, application-specific photolithography, but the underlying structure of all photo-processes is based on a number of common steps.

Wafer Preparation

A necessary requirement before photolithography commences is that the wafer be free from particle contaminants and has been desorbed of any moisture. Ensuring the latter is especially important because wafer cleaning is via a wet process ending in a DI water rinse and dry. A dehydrated surface can be achieved by baking at a temperature above 150 °C for several minutes prior to the application of photoresist. To complete the preparation, the wafer is coated with an adhesion promoter such as hexamethyldisilazane.

Photoresist Application

The wafer is immediately coated with liquid photoresist following preparation. The resist is dispensed on the centre of the wafer which is held via a vacuum seal on a metal or polymer chuck. When approximately 1–10 mL of resist has been dispensed, the wafer is spun at a typical speed of between 1 and 5 krpm. This distributes the resist over the entire wafer surface (Figure 5.6a).

Soft Bake

A post-spin soft bake is used to drive off most of the solvents in the resist while at the same time improving resist uniformity and adhesion. Typically, this is performed at 100 °C for a few minutes.

Exposure to Ultraviolet Light

The wafer is transferred to the mask-aligner where it is placed, with submicron precision, relative to the permanent pattern defined on the mask. Unless this is the first wafer layer, the pattern will be integrated with all previous layers. Once correctly aligned, the wafer is exposed to UV light (Figure 5.6b). In a positive resist process, the light passes through the transparent regions of the mask and activates the photosensitive components of the resist, such that these areas of resist are removed during the developing stage (see below). In a negative resist process the unexposed areas are removed.

Photoresist Developing

The photoresist pattern is created at the developing stage during which the wafer is exposed to a developing solution. Whether the process

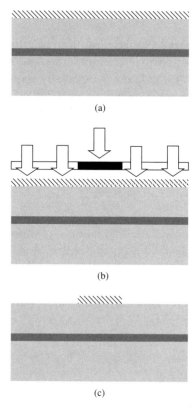

(a)

(b)

(c)

Figure 5.6 (a) The SOI wafer is uniformly coated with a thin polymer known as photoresist. (b) The resist is exposed to UV light through a permanent mask. The mask shown here is designed to result in waveguide formation. (c) Following hardbake, the desired pattern is printed in the photoresist ready for transfer to the wafer

is positive or negative, the solution dissolves the activated resist (or unactivated resist), leaving behind the resist pattern.

Hardbake

The final hardbake drives off the remaining resist solvents and further strengthens the resist adhesion to the wafer surface. It is typically carried out at a temperature of 90–140 °C for up to several minutes. Although the bake temperature may vary significantly depending on the next process step to be undertaken, the upper limit must be such that the hardbake does not result in pattern deformation via resist flow. Following hardbake, the desired pattern is printed in the photoresist (Figure 5.6c).

The photolithography process outlined in this section implies the use of the photoresist as the mask for the subsequent processing step. In some cases (for example silicon etching described below in section 5.2.2), it may be more appropriate to use a so-called 'hard mask' such as silicon dioxide or silicon nitride. If this is the case, the photolithography process is used as a mask for etching of the thin dielectric layer, which is subsequently used as a hard mask.

5.2.2 Silicon Etching

Etching refers to the controlled removal of material from the silicon wafer surface via a chemically reactive or physical process. There are two general approaches: wet and dry etching. Each approach has advantages and disadvantages, but for reproducing features of submicron dimensions dry etching dominates. Low-loss silicon waveguides, having dimensions typically >1 µm, have been produced by both wet [12] and dry [13] etching, but with the requirement for flexible process capability, tight tolerances and reproducible production, dry etching is regarded as the most suitable solution.

Dry etching proceeds through the formation of a low-pressure plasma. The physical processes which are involved in the formation and stabilisation of a plasma are complicated; however, because of the prevalence of plasmas in silicon processing (e.g. etching, deposition, photoresist removal, ion implantation) it is appropriate to explore some of the basic principles.

A plasma is an ionized gas, virtually neutral overall, consisting of electrons, ions and mostly neutral particles. The formation of a localised plasma (such as that required for processing) can be achieved by the application of either DC or AC power to a process gas contained in an isolated chamber, although DC bias is rarely used in practice because it is incompatible with insulating electrodes causing surface charging and therefore an unstable plasma. AC bias circumvents this problem with charge build-up in one half-cycle, followed by charge neutralisation in the next half. AC plasma generation therefore dominates in the design of semiconductor processing equipment, with the most common frequency being 13.56 MHz. A representation of a typical plasma setup is shown in Figure 5.7.

To begin the process, the plasma gas is introduced into the evacuated chamber and stabilised at a pressure between 10^{-3} and 10^{-2} Torr. Initial application of the AC signal causes free electrons in the gas to be accelerated greatly. These electrons lose energy through interaction

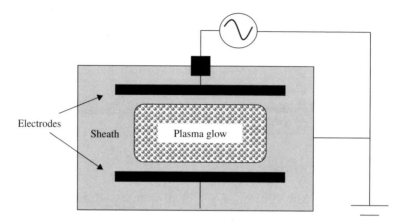

Figure 5.7 Schematic of a confined AC-generated plasma suitable for silicon processing. The processed wafer in placed on the lower, grounded electrode

with the gas atoms via processes such as molecular ionization and dissociation. The excited gas in turn gives rise to the emission of light and a distinctive glow is observed. The colour of the glow is dependent on the elemental constituents of the gas.

The vast disparity between the mass of the electrons and ions results in a time-averaged build-up of negative charge at the electrode surfaces, and a subsequent depletion of electrons in the gas volume close to the electrodes. Without electrons to drive the process, the excitation of the plasma gas in these volumes ceases and dark regions form (this is commonly referred to as a *sheath*). The average DC potential in the plasma chamber is determined by this distribution of charge. The result is an effective plasma potential (V_p), greater than the potential at either electrode, causing the acceleration of positively charged ions to the grounded electrode on which the process wafer is usually placed. The energy of ions is typically in the region of a few hundred electron-volts. Figure 5.8 shows the shape of the potential distribution between the electrodes.

A plasma used commonly in silicon processing is derived from CF_4 gas. CF_4 is normally stable, but dissociates (for example) into CF_3 and F atoms in a plasma with the single fluorine atom being used as the active etch element for both Si and SiO_2. A plasma based entirely on CF_4 would provide a slow etch rate owing to the swift recombination of CF_3 and F. By the addition of O_2 in the gas mix, the silicon etch rate can be dramatically increased owing to the reaction of oxygen with CF_3 inhibiting F recombination and hence increasing the free F

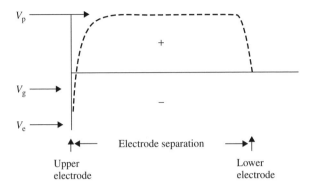

Figure 5.8   Representation of the time-averaged potential distribution in the plasma chamber. The labels 'upper' and 'lower' electrode refer to Figure 5.7. With the process wafer placed on the grounded (lower) electrode, singly charged ions arrive with an energy V_p

concentration. Fluorine-based chemical reactions have formed the basis of many different silicon etching processes [14].

This rather simplistic model of silicon dry-etching shields the reality of a set of complicated physical and chemical interactions. Plasma-based etchers are, however, ubiquitous in silicon processing and for a complete review of equipment and techniques the interested reader is referred to a more detailed text and references therein [15]. One process that is popular for etching features with effectively vertical walls (i.e. anisotropic etching), is *reactive ion etch* (RIE). RIE is a technique that uses both chemical (reactive) and physical (sputtering) processes to remove material from the wafer surface. Unlike the generic etch chamber described above and shown in Figure 5.7, the wafer is positioned on the AC-driven electrode, which is significantly smaller than the grounded electrode. In this way, a large voltage develops between the plasma and the wafer, ensuring the plasma-generated reactive ions have a degree of directionality normal to the wafer surface. This in turn ensures preferential etching in the direction of acceleration. An important consideration in any silicon etching process (both wet and dry), but particularly for RIE, is the need for selectivity to the masking material. In general the greater the degree of physical etching used, the greater the erosion of the masking medium (usually photoresist).

5.2.3   Critical Dimension Control

We have discussed some of the requirements for dimensional control previously in section 5.2.1 in relation to the demands on the

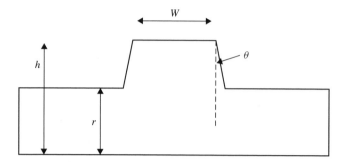

Figure 5.9 Schematic of a silicon waveguide. The dimensions critical to device performance are highlighted: rib width (W), silicon overlayer thickness (h), silicon thickness following rib etch (r) and rib wall angle (θ)

photolithography process. In fact, all of the steps in forming a silicon rib will effect a change in the dimensions of the rib, and each requires characterisation and constant monitoring. In high-volume manufacture this will entail the use of *statistical process control* (SPC).

The dimensions critical to waveguide performance have been described in detail by Powell [16], and are shown in Figure 5.9. The silicon overlayer thickness (h) is determined primarily by the process used to produce the SOI waveguiding layer. Silicon-on-SiO_2 lends itself particularly well to optical characterisation, such as that performed with an infrared reflectometer or ellipsometer, and hence material may be screened easily prior to the fabrication of silicon photonic circuitry [3].

The width of the rib (W) and the rib wall angle (θ) depend on both the photolithography and the silicon etch. In the case of W, a process bias will exist at each process step. For instance, the width of the rib image on the photo hard mask is generally different from that produced in the developed photoresist, and the ultimate width of the etched rib will be smaller than the photoresist image.

The rib height ($h - r$) is determined by the dry silicon etch. Optical absorption and interferometric techniques exist allowing the in-situ determination of etch depth with a considerable degree of accuracy – so-called *end-point detection*. Subsequent to etch a useful method for measuring an etch depth of micron dimensions is via surface profilometry.

Finally, an important consideration in calculating process bias for waveguides which have an upper oxide cladding is the reduction in h, r and W following thermal oxidation (see section 5.3). Because thermal oxidation proceeds via a reaction of oxygen and silicon atoms, the effect is a consumption of silicon and hence reduction of critical dimensions. A comparison of the atomic density of silicon ($5 \times 10^{22}\,cm^{-3}$) and the molecular density of SiO_2 ($2.2 \times 10^{22} \, cm^{-3}$) leads immediately to the fact that, for the growth of a given thickness t of SiO_2, a thickness of silicon equal to $(0.44 \, t)$ is consumed.

5.3 OXIDATION

The ability to easily form a high-quality, stable oxide is arguably the most important reason as to why silicon became the predominant material used in the microelectronics industry. SiO_2 has many applications in electrical IC fabrication, such as electrical isolation and implantation masking (see section 5.5).

The deposition or growth of a SiO_2 layer on an SOI wafer also converts the slab waveguide to a symmetrical from an asymmetrical waveguide. In order to provide sufficient optical confinement, and for the SiO_2 refractive index to dominate the cladding properties of the waveguide, it is necessary to prevent penetration of the electric field beyond the cladding layer. In the case of SiO_2 on silicon waveguides of several microns in dimension, the oxide thickness layer should be $>0.4 \, \mu m$ [17].

The most suitable method to achieve SiO_2 cladding layers on silicon waveguides is via thermal oxidation. This process not only guarantees a conformal, high-quality oxide suitable for integration with any electrical components on the same chip, but has the benefit of smoothing silicon surface roughness (either intrinsic or process-induced). Thermal SiO_2 growth proceeds via a relatively simple chemical process:

$$Si + O_2 \rightarrow SiO_2 \qquad (5.1)$$

for dry oxidation where the process gas is oxygen; and:

$$Si + 2H_2O \rightarrow SiO_2 + H_2 \qquad (5.2)$$

for wet oxidation where the process gas is steam. Silicon oxidation takes place in a quartz furnace in which the atmosphere and flow rate of the process gas are carefully controlled. Typical growth temperatures are in the range 750–1100 °C.

Thermal oxidation of silicon is well described by a model developed by Deal and Grove [18]. Their classic paper provides a general equation of the form:

$$2d_0/A = \{1 + [(t + \tau)/(A^2/4B)]\}^{0.5} - 1 \qquad (5.3)$$

where d_0 is the SiO_2 thickness, t is oxidation time, and A, B and τ are constants related to the reaction rate of oxygen at the silicon surface, the diffusion of the O_2 molecule in the oxide, and a shift in the time coordinate to account for the presence of an initial oxide, respectively. These three constants are determined experimentally for specific growth conditions. As an indication, for a dry oxidation at 1000 °C and a pressure of 760 Torr, values of $A = 0.165\,\mu m$, $B = 0.012\,\mu m^2/h$ and $\tau = 0.37$ hours give reasonable approximations to real oxide growth. Directly from equation 5.3 one can infer a reaction-rate limited process during the initial stages of the oxidation process where $(t + \tau) \ll (A^2/4B)$. This is related to the arrival of oxygen molecules to the wafer surface. For longer times, $t \gg \tau$ and $t \gg (A^2/4B)$, the process becomes limited by the diffusion of the oxygen or water molecules through the already present oxide layer before a surface reaction can take place.

Using the Deal and Grove model it is possible to generate plots of SiO_2 thickness versus annealing time (Figure 5.10). Most striking is the difference between wet and dry oxidation. At 1000 °C, the oxidation rate is approximately 20 times faster in a wet atmosphere. This is a direct result of the higher equilibrium concentration (by nearly three orders of magnitude) of H_2O compared to O_2 in SiO_2. Other factors affecting the oxidation rate are gas pressure, wafer crystal orientation, wafer dopant concentration, and wafer crystal damage (e.g. following ion implantation).

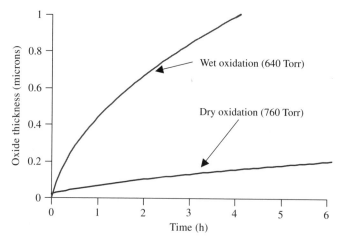

Figure 5.10 SiO_2 thickness versus oxidation time as derived from the Deal and Grove model [18]. The process temperature is 1000 °C

5.4 FORMATION OF SUBMICRON SILICON WAVEGUIDES

Throughout this chapter we have tacitly implied the application of the processes to silicon rib waveguides with a cross-section of several microns. However, there is considerable interest in the fabrication of submicron devices which permit high device packing density, low-cost production, and high-yield. As well as more conventional use in the telecommunication network, such small waveguides also have applications in on-chip communication and device interconnects.

Much of the process flow in fabricating low-dimensional waveguides is similar to that for the larger cross-section devices. However, as one might expect, the tolerances permitted in processing smaller devices scale with the device dimension, creating significant challenges for the process engineer. Silicon photonics can rely on processes developed in the microelectronics industry which has been working with deep-submicron devices for some considerable time.

The current state of silicon device processing is summarised each year in the *International Technology Roadmap for Semiconductors*, published by Sematech [10]. A study of the most recent roadmap indicates that the limitations of microelectronic production processes do not inhibit the formation of submicron waveguide devices. For instance, minimum CD control in photolithography is approximately 10 nm, with minimum feature sizes <100 nm. However, some unique issues do exist for small waveguide fabrication and these are brought to the attention of the reader in this section.

5.4.1 Silicon Dioxide Thickness

We have seen in Chapter 2 that an optical signal is not totally confined within the core of a waveguide. Hence, if it is desired that the index distribution of the guide be symmetrical in the vertical direction, it is necessary to ensure that the decay of the optical power into the cladding is essentially complete. In particular, if we require the cladding to be SiO_2, the buried oxide and the thermally grown surface oxide must be thick enough to prevent mode penetration into the silicon substrate and the over-oxide layer (in the case of a bare rib waveguide this would be air) respectively.

Mode penetration is a function of, *inter alia*, the thickness of the waveguide core, with more optical power propagating in the cladding as the size of the core is reduced. For large cross-section waveguides of multiple-micron dimensions, it has been found that an oxide thickness of 0.4 μm is adequate to provide good optical confinement [17]. As the size of the waveguide core is reduced to below a micron, the thickness of SiO$_2$ required to provide optical confinement increases to values beyond 1 μm (see Chapter 4).

The maximum thickness for the buried SiO$_2$ layer in SOI material fabricated using the SIMOX process is approximately 0.5 μm. This leaves wafer-bonded SOI as the most viable option for the fabrication of submicron waveguides. SOI formed by wafer bonding with a buried SiO$_2$ thickness up to 3 μm are commonly available. Of the two wafer-bonding techniques outlined in section 5.1, only UniBond® wafers possess the uniformity required for low-dimensional device fabrication. The range of silicon overlayer thickness that is available using the SmartCut® process (up to 1.5 μm) provides the additional benefit of starting material with no requirement for additional epitaxial growth.

Using thermal oxidation to produce a silicon surface oxide (upper cladding) of >1 μm is unreasonable, particularly in the case of low-dimensional silicon. It is likely that small waveguides will require CVD-deposited SiO$_2$ as an upper cladding layer. The principles of CVD were dealt with in 5.1.4 in relation to epitaxial silicon growth. High-quality conformal dielectric layers can be produced in an identical way with the use of a suitable processing precursor gas. Because CVD is a deposition process, there is no consumption of silicon and hence no reduction in silicon dimension.

5.4.2   Surface and Interface Roughness

Optical loss resulting from interface scattering was dealt with in section 4.6.1. Equation 4.21 shows that as the thickness of the silicon overlayer is reduced, interface scattering loss increases for equivalent micro-roughness. This phenomenon is of considerable concern to those wishing to fabricate low-dimensional silicon waveguides.

Interface roughness, measured as a roughness variance or r.m.s., is either intrinsic to the starting material, or is induced by a device processing step such as dry silicon etch. The silicon roadmap indicates surface micro-roughness for starting material of <0.2 nm and this holds for SOI wafers formed via either the SIMOX or the SmartCut® process. The interface roughness between the silicon overlayer and the buried SiO$_2$ has been reported to be below 1 nm for SIMOX [19], while UniBond® wafers, in principle, can exhibit interface roughness comparable to that found between a thermally grown SiO$_2$ on a silicon substrate. The

Figure 5.11 SOI rib formed via dry plasma etching. Etch-induced roughness is visible as the dark pitted regions each side of the rib. Reproduced by permission of P. Jessop, McMaster University

roughness of a silicon surface which has undergone a dry etch process may be increased by almost two orders of magnitude [20]. Figure 5.11 shows an SOI rib formed via a masked plasma etch process. The increase in surface roughness for the etched areas is clearly visible.

Thermal oxidation of silicon can be used as a method for the reduction of process-induced surface roughness. Specifically, it has been demonstrated that growing SiO_2 to a thickness >100 nm will ensure an r.m.s. of approximately 0.3 nm. This has been demonstrated for surfaces with initial roughness values of 3 nm, and may be appropriate for smoothing surfaces of greater roughness [21]. Of course, the use of thermal oxide for smoothing low-dimensional waveguides seems to oppose the deposition of thick CVD SiO_2 for waveguide upper cladding. However, the thermal oxide can be either sacrificial (i.e. removed by wet chemical etch before CVD deposition), or can form the initial layer of a thermal/CVD SiO_2 stack.

5.4.3 Sidewall Roughness

Using the same arguments as for the surface and interface roughness, the importance of the sidewall roughness of the silicon waveguide is highlighted. This precise issue was recently investigated by Lee et al. [22] for silicon waveguides with a height of 0.2 μm and widths varying from

0.5 to 8 μm. The ribs were formed using reactive ion etch and were bounded by a buried oxide of 1 μm and a deposited upper cladding oxide of 0.2 μm. A clear relationship between optical transmission loss and waveguide width was demonstrated and attributed to the increasing influence of sidewall roughness with reduction in waveguide dimension. The sidewall roughness of the etched waveguides was 9 nm needing to be reduced to 0.5 nm for the lowest dimensional waveguides to exhibit loss values of $<0.1\,\mathrm{dB\cdot cm^{-1}}$.

Lee and associates suggested the use of a wet chemical etch as opposed to dry etch to produce waveguides with low r.m.s. roughness. It is also possible that a thermally grown oxide would dramatically reduce the propagation loss by smoothing the sidewall roughness in a similar manner to that described in [21].

5.5 SILICON DOPING

Having described the basic process steps in the formation of the passive silicon waveguide, we now turn to the fabrication necessary to integrate electrical functionality with the optical silicon device. We will use, as an example, one of the simplest, but also most useful, electrical devices used in silicon photonic circuits: the $p-i-n$ diode [23]. This device is discussed in detail in section 6.1 of Chapter 6 and shown schematically in Figure 5.12. Designs of $p-i-n$ diodes rely on the fact that for low-loss waveguiding, the silicon overlayer must be pratically free of dopant material. However, by selective introduction of doped regions close to the waveguide (but far enough away to ensure no optical absorption), a monolithically integrated diode allows the controlled injection of carriers

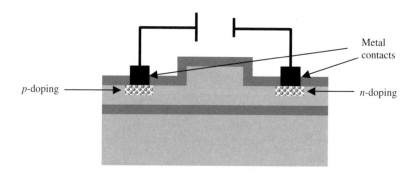

Figure 5.12 Silicon waveguide with a monolithically integrated $p-i-n$ diode. The diode is shown forward-biased and as such will induce a phase change and simultaneous increase in optical attenuation

such that they interact with the optical signal under the rib. We saw in section 4.8.2 that both phase shifting and absorption can be controlled in this way.

5.5.1   Ion Implantation

In section 4.6.2 it was demonstrated that for low-loss propagation of 1.55 μm light in silicon the concentration of electrically active impurities such as phosphorus and boron must be $<1 \times 10^{16}\,cm^{-3}$. This corresponds to a resistivity value $>10\,\Omega\cdot cm$. Wafers are readily available at resistivities of several tens of ohm·cm, including SOI fabricated via SIMOX and SmartCut®. Silicon epitaxial growth produces layers that are practically intrinsic (resistivity of hundreds of ohm·cm).

To produce electrical and optical functionality in the same device and hence create a silicon photonic integrated circuit we are required to selectively reduce the resistivity of the silicon wafer. This is achieved by the introduction of group III atoms (e.g. boron for *p*-type behaviour) and group V atoms (e.g. arsenic or phosphorus for *n*-type behaviour) into the silicon crystal lattice, in a process known as *electrical doping*. As a consequence, the group III and V impurities are referred to as *dopants*. For successful electrical device fabrication the concentration and distribution of the dopants must be tightly controlled. Importantly, the introduction of the dopant must be restricted to volumes that do not form any significant part of the optical waveguide (i.e. there must be minimal unintentional, passive interaction of the dopant and the optical signal).

During the first few decades of semiconductor device fabrication doping was performed via the deposition onto, and subsequent diffusion of impurity into, a silicon wafer in a raised temperature furnace rather like that used for thermal SiO_2 growth. Typical source gases might be arsine (AsH_3) or diborane (B_2H_6). The volume of silicon doped was controlled by the use of windows etched into overlayers of SiO_2, so called *masking* (the diffusion of dopants in oxide being several of orders of magnitude less than in silicon). To ensure a repeatable process the concentration of the dopant was determined in the deposition stage by its solid solubility. This severely restricted the concentration profiles and range in device performance that could be obtained.

In the 1970s, ion implantation was developed as an alternative doping technology. Its main advantages are related to precise control of the delivered dopant purity and dose and flexibility in the concentration and profile that can be achieved. Further, because implantation is a

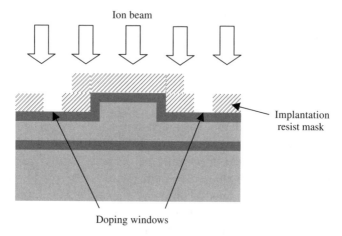

Figure 5.13 For a real selective implantation, doping windows are opened in an implantation mask allowing the passage of the ion beam into the silicon device

low-temperature process it is compatible with masks with poor thermal stability, such as photoresist, removing the need to mask the doping process with a SiO_2 overlayer. A photoresist mask suitable for the production of a p–i–n diode integrated with a silicon rib is shown in Figure 5.13. In this example, a thermal SiO_2 upper cladding is in place and the ions are selectively implanted through the oxide and directly into the silicon wafer.

5.5.2 The Implantation System

A significant drawback of ion implantation is the complexity and relative high cost of ownership of the implantation system, although these are outweighed by its ability to deliver precisely the vast range of doping profiles required for the fabrication of the most complex device structures. An ion implanter has four major subsystems: (1) ion source, (2) magnetic analyser, (3) accelerator, and (4) process chamber, shown schematically in Figure 5.14.

Doping by ion implantation, not surprisingly, requires the dopant impurities to be in the form of ions. Source material such as boron trifluoride gas or BF_3 is broken down into a variety of species including BF_2^+, BF^+, F^+ and B^+ in the ion source. This is achieved by the creation of localized plasma driven by the acceleration of electrons provided by a tungsten filament. All positive ion species generated in the source are extracted through a small aperture, and this beam of charged particles is subjected to mass selection using an analysing magnetic.

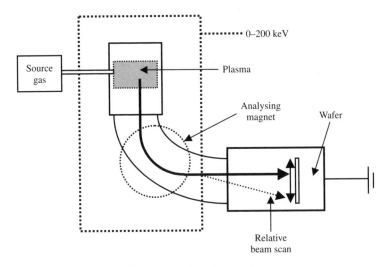

Figure 5.14   Schematic of an ion implantation system

The magnetic field (**B**) is positioned perpendicular to the direction of beam propagation. Under the influence of **B**, the ions that make up the beam are deflected to a degree dependent on their velocity, and mass. The heavier ions, such as BF_2^+, are deflected less than the lighter ions, such as B^+. With careful adjustment of **B**, ions of any particular mass can be selected from the analysing magnetic subsystem through an aperture, and into the acceleration subsystem.

Upon entering the accelerator, the desired ions are subjected to an electric field, the magnitude of which determines their final energy and hence ultimate depth in the silicon wafer. Various machine designs exist which extend the range of ion energies from 200 keV to >1 MeV, but the most common tools reliably implant ions with energies ranging from sub-keV to a maximum of 200 keV.

The process chamber houses the silicon wafer to be implanted. Uniform exposure of the wafer to the ion beam, which may only be a few square centimetres in diameter, requires relative beam scanning. There are two approaches: mechanical scanning of the wafer with a stationary beam, or electrostatic scanning of the beam with the wafer remaining stationary. The former is predominantly used for high-current, high-throughput machines, while the latter is commonly found in low- to medium-current, single-wafer tools. In-situ dose measurement is performed in real time by allowing the beam to be overscanned (i.e. beyond the diameter of the wafer) and impinge onto a Faraday cup where a representative beam current is measured. The beam dosimetery may be calibrated later by measuring the resistivity of the implanted wafer.

5.5.3   Implantation Parameters

The ion dose (Q) is the number of ions implanted per unit area (usually measured in cm^{-2}) and is calculated using:

$$Q = I\,t/e\,n\,A \qquad (5.4)$$

where I is the measured current, t is the time of implant, e is the electronic charge, n is the charge state of the ion (for B^+, $n = 1$), and A is the area of the charge collector in square centimetres. The ability to monitor or measure I, t and A to an accuracy of <1 % leads directly to repeatable dose control for the implantation process.

Each implanted ion will impinge on the wafer with the same energy. However, the subsequent slowing down process is randomly dependent on ion interactions with the constituent free electrons and silicon atoms (discussed in more detail below). The average path length of the implanted ions is called the *range* (R). R has components in both the vertical and horizontal directions. Of significance to device designers is the average distance travelled in the direction perpendicular to the wafer surface, and this is called the *projected range* (R_p), with the thickness of the implanted distribution called the *ion straggle* or ΔR_p. The projected range is a function of the ion energy and mass. The distribution of implanted ions can be approximated analytically by a gaussian, while more complex statistical distributions give better approximations. Numerical solutions for the ion distribution can be obtained also from Monte Carlo simulations and one of the most popular is the freely available TRansport of Ions in Matter (TRIM) code [24]. Figure 5.15 shows plots derived from TRIM simulations for B^+ ions implanted into silicon for energies of 50, 100 and 150 keV, to a total dose for each energy of 1×10^{15} cm^{-2}. It is clear that as the implantation energy is increased both the projected range and ΔR_p increases. Also, the projected ranges associated with these typical energies are between 0.15 and 0.6 μm.

The slowing down of the energetic ions in the silicon can be described by two general mechanisms: electronic stopping and nuclear stopping. *Electronic stopping* occurs as the ions interact with the target electrons and creates no permanent damage in the silicon wafer. However, it will cause the ejection of electrons when they are situated within a few tens of nanometers of the wafer surface. This can result in severe wafer charging especially when the substrate is insulted from the outside world (as is the case for SOI wafers). Charging effects can be eliminated during the implantation process by providing a steady supply of low-energy electrons to the wafer surface extracted from a nearby remote gas plasma.

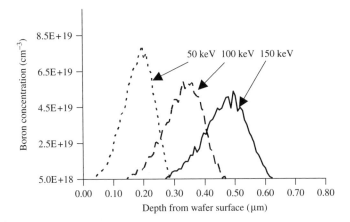

Figure 5.15 Monte Carlo simulation [24] of the distribution of boron ions following implantation into silicon for 50, 100 and 150 keV implantation energy. The dose for each energy is 1×10^{15} cm^{-2}

Nuclear stopping describes the collision of the implanted ions with the silicon atoms. These collisions cause wafer atoms to be displaced from their lattice site creating a lattice vacancy and a silicon interstitial (i.e. an atom which resides off the crystalline lattice). Although many of these defects 'repair' themselves at room temperature, approximately 5 % of the primary vacancies and interstitials form more complex defects that are stable at room temperature [25].

5.5.4 Dopant Activation and Drive-in

A high-temperature anneal (>900 °C) will reconstruct the silicon lattice to its crystalline condition and simultaneously position the dopant ions onto lattice sites. This latter process is known as *electrical activation* and is the final step in the doping process. The anneal also causes the dopant atoms to diffuse. This can be useful if the dopant distribution is required to be broader, and the R_p deeper, than is available from the as-implanted profile. This post-implant redistribution of dopant is referred to as 'drive-in' and typically requires a furnace anneal at temperatures in excess of 1000 °C for many minutes. Alternatively, there may be a requirement for the profile of the as-implanted ions to be retained as intact as possible. In this case, the electrical activation is achieved by a so-called 'rapid thermal anneal' (>1000 °C for a few seconds).

In the formation of the *p–i–n* structure shown in Figure 5.12 and the device structures described in section 6.1.3 of Chapter 6, the *p*-type and *n*-type regions are heavily doped in a similar manner to the source and drain contacts in CMOS technology (i.e. $>1 \times 10^{15}\,cm^{-2}$). For doped regions on each side of the rib, it is necessary to prevent the attenuation of the optical mode in the device 'off' state by opening the implant windows far enough away from the rib walls. In addition, consideration must be made for lateral thermal diffusion during the high-temperature anneal. For doped regions in the top of the rib (such as that shown in Figure 6.2), the electrical activation should proceed via a rapid thermal anneal to prevent any significant diffusion of the implanted dopant.

5.6 METALLISATION

The final step to the formation of an integrated silicon photonic device requires the pattern definition of metal on the silicon wafer for device interconnects and access to the device from the outside world. Similar caution is required with metallisation as with doping, in that none of the metal and any part of the optical mode should interact, or else significant absorption will result. In practical terms this demands sufficient upper-cladding oxide thickness to prevent optical power decay beyond the oxide and into the metal layer.

5.6.1 Via Formation

The resistivity of SiO_2 is extremely high (approximately $10^{15}\,\Omega\cdot cm$) and hence the upper-cladding SiO_2 provides excellent electrical insulation between the metal/silicon and metal/metal in the case of multilevel metallisation. If we again refer to our simple $p-i-n$ structure in Figure 5.12, we observe the requirement for intimate contact only between the metal and the heavily doped p and n regions and hence the selective removal of SiO_2 from the doped silicon surface. Metal contact with undoped regions of the wafer surface is undesirable as it will result in current leakage and parasitic capacitance effects. Using photolithography, photoresist windows are formed inside the doped area. The SiO_2 is then etched away from the wafer surface using a hydrofluoric-acid (HF) based wet etch. The resulting SiO_2 windows are referred to as contact 'vias'. Metal film deposition is performed within quick succession of the removal of the surface oxide. This ensures the minimum regrowth of 'native' SiO_2 following exposure to the atmosphere.

5.6.2 Metal Deposition

A traditional method for depositing metal which is still used in small-scale research and development is *thermal evaporation*. The metal to be deposited is placed in a crucible inside a vacuum chamber. The chamber is evacuated to around 10^{-6} Torr and the metal source is heated by an electron beam until vaporization occurs. The wafer is placed far enough away from the source to ensure a uniform deposited thickness across the wafer area. Thermal evaporation provides a robust and inexpensive method for silicon metallisation, but significant drawbacks are its inability to provide uniform step coverage of etched features and the requirement for each metal type to have a dedicated crucible.

A more common method for metal deposition is *sputtering*. The sputtering technique users energetic, inert gas ions to strike a source-metal target and physically dislodge metallic atoms – rather like a cannon-ball striking a muddy field. The sputtered atoms are allowed to migrate through a vacuum until they are deposited on a target wafer. The inert gas (usually argon) is ionized inside an evacuated plasma chamber in a similar process to that described in section 5.2.2. The use of a 13.56-MHz power source has resulted in this technique being referred to as *RF sputtering*. Figure 5.16 shows a schematic representation of an RF sputtering chamber. The RF bias is applied to the back surface of the metal target. The formation of plasma and creation of the plasma potential attracts positive argon ions to the target where they dislodge target atoms. These atoms then migrate towards the wafer surface where they form the metal layer.

RF sputtering suffers from a relatively poor sputtering yield and hence slow deposition rate. Using magnetic confinement (magnetron sputtering), electrons close to the target are prevented from escaping

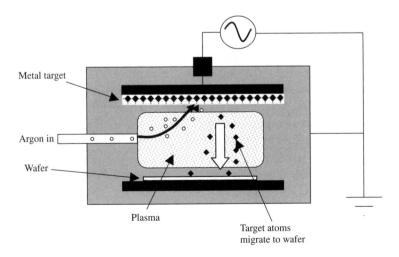

Figure 5.16   Schematic of an RF sputtering system

from the discharge region. These electrons instead contribute to the ionization rate of the plasma, and subsequent bombardment of the target metal, significantly increasing the sputter yield and deposition rate.

5.6.3 Material Choice

The choice of metal to be deposited depends on many parameters, such as available deposition tools, the specific application of the metallic layer and the reliability requirements for the device. In general though, semiconductor/metal contacts are required to have negligible resistance compared to the device itself, so-called 'ohmic' contacts, while interconnects between devices must have as low a resistance as possible. The traditional 'work-horse' metal of the microelectronics industry is aluminium. It has low room-temperature resistivity (2.5 $\mu\Omega\cdot$cm), is relatively inexpensive, is easily deposited, and reacts with both silicon and SiO_2 ensuring excellent adhesion to device layers. In addition, aluminium can be patterned and etched in a straightforward manner to form the complex interconnects and contacts required in IC fabrication (for etching recipes see [2]). Although aluminium is reaching the limits of use in deep submicron technology, it is worth mentioning that for prototype device development in academic or small-scale industrial operations, and devices where shallow doped regions are not required, aluminium is still an excellent choice for general-purpose metallisation.

In recent years copper has emerged as the replacement for aluminium in deep submicron interconnect fabrication. The main reason for this is its resistivity of only 1.7 $\mu\Omega\cdot$cm increasing speed while reducing the power required to drive a device (alternatively increasing the packaging density for the same power). Processing with copper is not straightforward. It diffuses quickly in Si and SiO_2, is resistant to many wet and dry etch processes making pattern definition difficult, and unlike aluminium it oxidises when exposed to atmospheric ambient.

Refractory metals (such as titanium, cobalt and molybdenum) react with silicon to form silicide when heated to temperatures ranging from 900 to 1400 °C, depending on the silicide material. The resulting compound has significantly greater thermal stability compared to aluminium. With a resistivity of tens of μohm·cm, silicide is suitable for forming the contact between metal layers and the silicon surface in modern submicron devices.

5.6.4 Sintering and Barrier Materials

The formation of a robust ohmic contact between aluminium and silicon necessitates heating the interface to a temperature of around 500–550 °C for several minutes in an inert or reducing ambient, in a process known as *sintering*. However, the use of pure aluminium can result in significant silicon diffusion into the aluminium layer during the sintering step. The resulting interface structure resembles the formation of aluminium spikes into the underlying silicon which may penetrate the doped junction (particularly if it is submicron in depth). This phenomenon is known as 'junction spiking'.

The most reliable method for the elimination of junction spiking is the incorporation of a diffusion barrier layer between the aluminium and the silicon. This thin layer must be able to prevent significant diffusion at the sinter temperature, have low electrical resistivity and possess good adhesion properties to both the Si and Al (or whatever metal is being used as a contact). Two of the most common barrier layers are titanium tungsten (TiW) and titanium nitride (TiN), although a thin layer of $TiSi_2$ is required below TiN to reduce the contact resistance.

5.7 SUMMARY

The development of silicon photonic fabrication benefits greatly from the vast library of knowledge that already exists in the silicon microelectronics industry. In this chapter we have described the basic steps in the formation of an integrated silicon photonic device and acknowledge that we have merely glimpsed at the general area of silicon fabrication. Our emphasis has been skewed towards fabrication in academic and research environments where much of the near-term device development will be performed. We concede that as with any fast-moving technology, silicon photonic processing will find a path of least resistance which may make some of the processes described here redundant and necessitate the use of methods not described (for instance in the inevitable development of submicron devices). However, the overall process flow and much of the detail of this chapter will remain relevant for many years to come.

REFERENCES

1. M Quirk and J Serda (2001) *Semiconductor Manufacturing Technology*, Prentice-Hall, New Jersey.
2. S M Sze (1988) *VLSI Technology*, McGraw-Hill, Singapore.
3. G K Celler and S Cristoloveanu (2003) 'Frontiers of silicon-on-insulator', *J. Appl. Phys.*, **93**, 4955–4978.
4. J Schmidtchen, A Splett, B Schüppert and K Petermann (1991) 'Low-loss single-mode optical waveguides with large cross-section in silicon-on-insulator', *Electron. Lett.*, **27**, 1486–1487.

5. W P Maszara, G Goetz, A Caviglia and J B McKitterick (1988) 'Bonding of silicon wafers for silicon-on-insulator', *J. Appl. Phys.*, **93**, 4943–4950.

6. A F Evans, D G Hall and W P Maszara (1991) 'Propagation loss measurements in silicon-on-insulator waveguides formed by the bond-and-etchback process', *Appl. Phys. Lett.*, **59**, 1667–1669.

7. M Bruel (1995) 'Silicon on insulator material technology', *Electron. Lett.*, **31**, 1201–1202.

8. M Bruel, B Aspar and A-J Auberton-Hervé (1997) 'Smart-Cut: A new silicon on insulator material technology based on hydrogen implantation and wafer bonding', *Jpn J. Appl. Phys.*, **36**, 1636–1641.

9. T W Ang, G T Reed, A Vonsovici et al. (1999) '0.15 dB/cm loss in Unibond SOI waveguides', *Electron. Lett.*, **35**, 977–978.

10. *The International Technology Roadmap for Semiconductors*, Sematech (2001).

11. R A Soref, J Schmidtchen and K Pesermann (1991) 'Large single-mode rib waveguides in Ge-Si and Si-on-SiO$_2$', *IEEE J. Quant. Electron.*, **27**, 1971–1974.

12. U Fischer, T Zinke, J-R Kropp, F Andt and K Petermann (1996) '0.1 dB/cm waveguide losses in single-mode SOI rib waveguids', *IEEE Photon. Technol. Lett.*, **8**, 647–648.

13. B Jalali, P D Trinh, S Yegnanarayanan and F Coppinger (1996) 'Guided-wave optics in silicon-on-insulator technology', *IEE Proc. Optoelectron.*, **143**, 307–311.

14. A M Voshchenkov (1993) 'Plasma etching: an enabling technology for gigahertz silicon integrated circuits', *J. Vac. Sci. Technol. A*, **11**(4), 1211–1220.

15. C Y Chang and S M Sze (1996) *ULSI Technology*, McGraw-Hill, Singapore.

16. O Powell (2002) 'Single-mode conditions for silicon rib waveguides', *J. Lightwave Technol.*, **26**, 1851–1855.

17. A Rickman, G T Reed, B L Weiss and F Namavar (1992) 'Low-loss planar optical waveguides fabricated in SIMOX material', *IEEE Photonics Technol. Lett.*, **6**, 633–635.

18. B E Deal and A S Grove (1965) 'General relationship for the thermal oxidation of silicon', *J. Appl. Phys.*, **36**, 3770–3778.

19. C Jacobs, A Genis, L P Allen and P Roitman (1994) 'Effect of anneal temperature on Si/buried oxide interface roughness of SIMOX', *Proc. IEEE Int. SOI Conf.*, pp. 49–50.

20. K-S Chen, A Ayón and S M Spearing (2002) 'Effect of process parameters on the surface morphology and mechanical performance of silicon structures after deep reactive ion etching (DRIE)', *J. Micromech. Syst.*, **11**, 264–275.

21. L Lai and E A Irene (1999) 'Limiting Si/SiO$_2$ interface roughness resulting from thermal oxidation', *J. Appl. Phys.*, **86**, 1729–1735.

22. K L Lee, D R Lim, H-C Luan et al. (2000) 'Effect of size and roughness on light transmission in a Si/SiO$_2$ waveguide: experiment and model', *Appl. Phys. Lett.*, **77**, 1617–1619.

23. P D Hewitt and G T Reed (2000) 'Improving the response of optical phase modulators in SOI by computer simulation', *J. Lightwave Technol.*, **18**, 443–450.

24. J F Ziegler, J P Biersack and U Littmark (1985) *The Stopping and Range of Ions in Solids*, Pergamon, New York.

25. A P Knights, F Malik and P G Coleman (1999) 'The equivalence of vacancy-type damage in ion-implanted Si seen by positron annihilation spectroscopy', *Appl. Phys. Lett.*, **75**, 466–468.

6

A Selection of Photonic Devices

This chapter comprises several sections, each discussing a particular photonic device. The devices are not all specific to silicon photonics, in the sense that they could be fabricated in many integrated optical technologies. However, it is helpful in some cases to refer to a specific waveguide technology when explaining some devices, and when this is required the silicon-on-insulator (SOI) technology is used as an example, although generality is retained if more appropriate.

6.1 OPTICAL PHASE MODULATORS AND VARIABLE OPTICAL ATTENUATORS

One of the requirements of an integrated optical technology, particularly one related to communications, is the ability to perform optical modulation. This implies a change in the optical field due to some applied signal, typically – although not exclusively – an electrical signal. The change in the optical field is usually derived from a change in refractive index of the material involved, or via a direct change in intensity of the optical wave, although other parametric changes are possible. It is now widely accepted that the most efficient means of implementing optical modulation in silicon via an electrical signal is to use carrier injection or depletion. This effect can be used to produce either *variable optical attenuators* (VOAs) or *phase modulators*, and so it is convenient to discuss these devices in a single section of this text. Firstly optical phase modulators will be discussed, and subsequently VOAs.

Silicon Photonics: An Introduction   Graham T. Reed and Andrew P. Knights
© 2004 John Wiley & Sons, Ltd   ISBN: 0-470-87034-6

This first section of the chapter also demonstrates how some of the material presented so far in this text can be incorporated into the design of a real device. Clearly there are numerous devices that could have been chosen, but an optical phase modulator (and/or a VOA) has been chosen because it is an active device, because phase modulators operate best as single-mode devices, and because both devices may sometimes involve some micromachining in the fabrication process. Furthermore these devices are central to the continued interest in silicon optical circuits.

6.1.1   The Optical Phase Modulator

The device discussed in this section was first designed at the University of Surrey [1], and fabricated in a University of Surrey/University of Southampton collaboration [2].

An optical phase modulator is a device that can dynamically vary the phase of an optical wave in a manner determined by some applied driving function. For the purposes of this text we will consider an optical phase modulator that operates via injection of free carriers, into the region in which an optical mode is propagating. The injection of carriers is therefore related to the current flowing in the device, which is usually regarded as the driving function.

6.1.2   Modelling of Semiconductor Devices

There are several detailed semiconductor device modelling packages commercially available. Most are obviously aimed at the electronic device industry, and therefore care must be taken when using such modelling tools for unusual applications. The modelling of an optical phase modulator in silicon is such an example, because not only is the device larger than most semiconductor devices, but it also operates at high levels of injected carriers, which is unusual for many semiconductor devices. In this work two commercial simulators were used, one called MEDICI, produced by TMA Inc., and a second called Silvaco, from Silvaco International. Both simulators produced near identical results, so the results given here are not attributed to one simulator or the other.

Both simulators operate in the same manner, solving the equations governing charge behaviour in semiconductor devices. These are principally the *continuity equation* and *Poisson's equation*. The former describes the evolution of charge transport, and the latter relates changes in potential to local charge densities. There are numerous supporting models that are used for variations in device parameters, but it is beyond the scope of the text to discuss the simulators in detail. However, we should note that the complex nature of carrier transport and distribution requires a range of models to adequately describe them sufficiently well, and hence care must be taken to ensure that both sufficient detail is included in the supporting models, and that constants are defined sufficiently accurately for the work being undertaken.

In common with most numerical models, a mesh is developed representing the device, and the defining equations are solved for adjacent mesh points. Consequently including a sufficiently detailed mesh for the device in question is important. However, a dense mesh will require more computing time to solve, and sometimes a less dense mesh can provide adequate results. Therefore, it is usual to design a variable mesh with areas of (spatially) more detail, where it is important to know more about the device behaviour.

6.1.3   Basic Device Geometry, and the Aim of Modelling

The purpose of embarking upon detailed modelling of a device is usually to maximise the performance of the device given certain constraints, prior to fabrication. In order to perform such an operation, it is necessary to decide in advance upon the priorities in terms of device performance. A common consideration is whether operating speed or electrical power consumption is more important. Alternatively, minimal optical loss may be the prime consideration. For the purposes of this text, we will consider a device in which we wish to maximise the DC performance of a phase modulator, because it is perhaps the simplest parameter for us to consider, whilst demonstrating the effectiveness of device modelling. In practice this means we wish to maximise the refractive index change for a given current. Alternatively, considering the charges within the device, this means we would like to have high carrier density only where the optical mode exists, and none elsewhere. This is an impossibility, so our task is to minimise the current flow (and hence the carrier density), in parts of the waveguide where little optical power propagates.

In order to achieve carrier injection in a semiconductor, we clearly need to set up current flow through the device. One of the simplest ways of doing this is to use a $p-n$ junction. However, since a $p-n$ junction requires relatively high doping concentrations in the p and n regions, such doping will cause high optical loss via absorption if the doped regions are formed in the parts of the waveguide where the optical power is concentrated. Consequently a $p-i-n$ structure is more

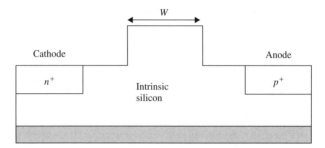

Figure 6.1   Optical phase modulator $p-i-n$ structure

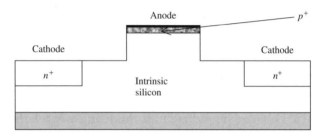

Figure 6.2   Symmetrical optical phase modulator $p-i-n$ structure

appropriate for such a device, the intrinsic region coinciding with the region of the waveguide containing the optical power. Therefore a series of $p-i-n$ structures can be imagined. Perhaps the simplest is the $p-i-n$ structure formed laterally across a rib waveguide, as shown in Figure 6.1.

However, in order to optimise the device for refractive index change, we need to achieve controlled current flow throughout the rib. Intuitively it seems likely that we could obtain better injection densities in the centre of the rib using an alternative $p-i-n$ orientation, such as that of Figure 6.2. This turns out to be the case, although that point is not specifically verified here. Let us consider the generalised modulator of Figure 6.2 and aim to maximise the refractive index change in this structure, for a given current flowing through the device. There are a number of geometrical parameters that we can consider:

1   The angle of the rib walls (shown vertical in Figure 6.2)
2   The rib width and height
3   Depth of the n^+ and p^+ regions
4   Lateral displacement of n^+ regions from the base of the rib.

The way in which we will consider the improvement in the modulator performance is to consider the predicted behaviour of the device with each parametric variation. More specifically we will consider the predicted density of injected carriers in the region of the optical mode. In order to do this accurately it is necessary to consider the overlap of injected carriers (charge) with the propagating optical mode. However, before we do this, if we investigate the basic structure of an optical phase modulator of the type shown in Figure 6.2, it becomes clear that we can simplify the process significantly. To do this let us begin with a device of specific dimensions.

Consider Figure 6.3. The device is based upon an SOI wafer with a silicon overlayer of 6.5 μm. In some senses this layer thickness is arbitrary, but the intention is that a large, multi-micron device is to be modelled, such that coupling to or from optical fibres is relatively efficient. Notice that the dimensions are such that the single-mode criteria as defined by equation 4.12 are satisfied.

Consider the device of Figure 6.3 under forward bias (i.e. a positive voltage applied to the p^+ terminal with respect to the n^+ terminals). This will cause holes and electrons to be injected into the intrinsic silicon region, constituting a current. Since the structure forms a $p–i–n$ diode it will have the diode characteristic associated with silicon devices, with a switch-on voltage of approximately 0.6 V. Therefore we must forward-bias the device by more than 0.6 V to obtain current flow. Consider for example, a forward bias of 0.9 V. The carrier injection concentration will be influenced by the doping concentrations of the p^+ and n^+ regions,

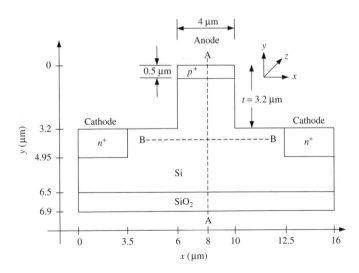

Figure 6.3 Generalised three-terminal phase modulator for performance modelling. Modified from Figure 3 in C K Tang, G T Reed, A J Walton, and A G Rickman (1994) 'Low loss, single mode optical phase modulator in SIMOX material', *J. Lightwave Technol.*, **12**, 1396. By permission of IEEE

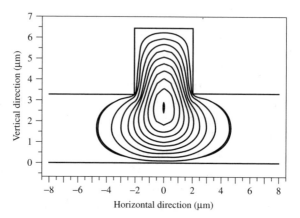

Figure 6.4 Profile of the fundamental mode in an SOI rib waveguide

which thus far are undefined. Initially let both regions be doped to a density of 5×10^{18} cm^{-3}. It is important to consider the concentration of carriers in the region of the optical mode. Clearly to do this we must model the mode profile. A typical mode profile is shown in Figure 6.4, with contours of optical intensity in 10 % steps. The exact mode profile will vary with changes in the rib geometry, but from Figure 6.4 we can see that even for a relatively shallow rib, almost 80 % of the power is contained under the rib.

Therefore, consider the profile of injected carriers into the device of Figure 6.3. With reference to the figure, we can consider the profile of injected carriers along the lines AA and BB, shown in Figures 6.5 and 6.6 respectively. We can see from both figures that the level of injected carriers is of the order of 2×10^{17} cm^{-3}, and the variation in injected carrier density across the region of the optical mode is of the order of 2×10^{16} cm^{-3}. Therefore, the variation in the injected level is an order of magnitude lower than the predicted absolute level of injection. Consequently, for the purposes of this section of the text, we will regard the level of injected carriers as approximately constant, certainly across the high-intensity regions of the optical mode. This avoids the need to evaluate the overlap integral between the mode profile and the injected carrier profile, and hence simplifies our modelling, as we can regard the refractive index change as approximately constant.

6.1.4 Effect of Parametric Variation on the DC Efficiency of an Optical Modulator

We will now consider the effect of the parametric variations listed in section 6.1.3, that were:

1 The angle of the rib walls (shown vertical in Figure 6.2)

2 The rib height and width

3 Depth of the n^+ and p^+ regions

4 Lateral displacement of n^+ regions from the base of the rib.

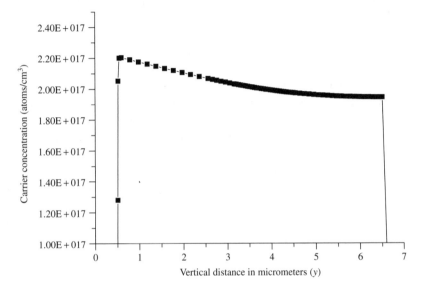

Figure 6.5   Injected carrier density along a vertical section at $x = 8$

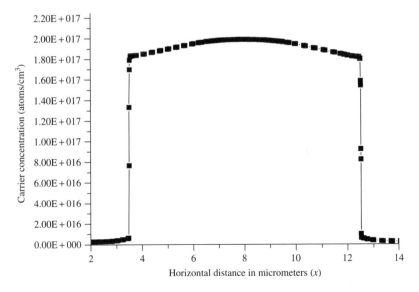

Figure 6.6   Injected carrier density along a horizontal section at $y = 3.8$

The Angle of the Rib Walls

Clearly it is possible to define a theoretical optical modulator with a rib wall at any angle to the vertical. However, to fabricate a device with a rib wall at a specific angle is very difficult. It is possible to fabricate a device with specific angles using the micromachining techniques discussed in section 4.9. Using an anisotropic etch we can achieve an angle of 54.7° to the surface (an exposed (111) plane). Consequently, let us compare the injected carriers for a device with vertical walls and one with angled walls (54.7°). The device is shown in Figure 6.7.

For the device in Figure 6.7, the parameters are the same as for the device in Figure 6.3, other than the angled walls. The two devices were simulated using the same simulator, and the results are shown in Figures 6.8 and 6.9. Once again, carrier densities are plotted in the y and x directions, along the lines AA and BB shown in Figure 6.7.

We can see from both Figure 6.8 and Figure 6.9 that the modulator with angled walls provides significantly improved carrier injection, as compared to the corresponding injection from the modulator with vertical rib walls. It is also worth noting that the uniformity of the carrier density is slightly increased for the angled rib walls.

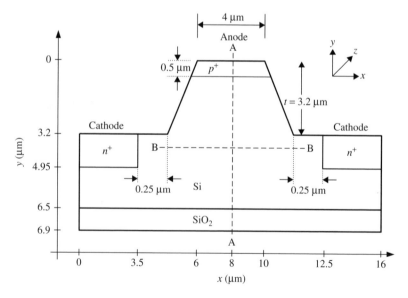

Figure 6.7 Three-terminal phase modulator with angled rib walls. Modified from Figure 2 in C K Tang, G T Reed, A J Walton, and A G Rickman (1994) 'Low loss, single mode optical phase modulator in SIMOX material', *J. Lightwave Technol.*, **12**, 1396. Reproduced by permission of IEEE

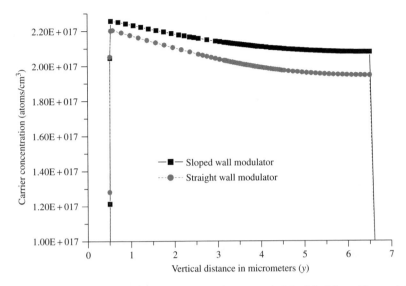

Figure 6.8 Comparison of injected carriers along $x = 8$. Modified from Figure 4 in C K Tang, G T Reed, A J Walton, and A G Rickman (1994) 'Low loss, single mode optical phase modulator in SIMOX material', *J. Lightwave Technol.*, **12**, 1397. Reproduced by permission of IEEE

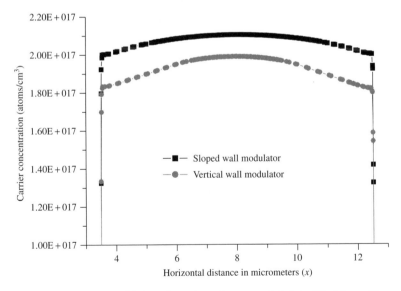

Figure 6.9 Comparison of injected carriers along $y = 3.8$. Modified from Figure 5 in C K Tang, G T Reed, A J Walton, and A G Rickman (1994) 'Low loss, single mode optical phase modulator in SIMOX material', *J. Lightwave Technol.*, **12**, 1397. Reproduced by permission of IEEE

The Effect of the Rib Height and Width

Let us now consider how variations in the rib height and width affect the injected carrier density. Firstly let us consider the rib height. By the rib height we mean the parameter labelled '$t = 3.2\,\mu$m' in Figure 6.7. In order to fulfill the single-mode requirements of section 4.4, t needs to be less than or equal to 3.25 μm. Therefore we can reduce the rib height and retain single-mode operation, if we satisfy the remaining conditions of section 4.4. Clearly there are many values that t could take. In order to demonstrate the effects let us assume a value of $t = 2\,\mu$m. We can now compare the injected carrier density for the sloped-wall modulator for the two values of t. The results are shown in Figures 6.10 and 6.11 with an expanded vertical scale.

Figure 6.10 shows that the injected carrier density along $x = 8$ is not greatly affected by changes to the rib height, particularly in the most intense part of the optical mode. If we look at the distribution in the y direction, the largest difference occurs towards the bottom of the waveguide. Figure 6.11 shows the variation along $y = 5.5$. Even here, where the variation is at its largest, it is still less than 1 %.

The variation of rib width may be expected to yield more dramatic changes in injected carrier density. Since the carrier profiles don't vary significantly with rib width (although of course the current density will

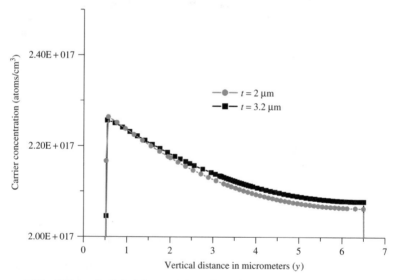

Figure 6.10 Effect of rib height on injected carriers along $x = 8$. Modified from Figure 6 in C K Tang, G T Reed, A J Walton, and A G Rickman (1994) 'Low loss, single mode optical phase modulator in SIMOX material', *J. Lightwave Technol.*, **12**, 1397. Reproduced by permission of IEEE

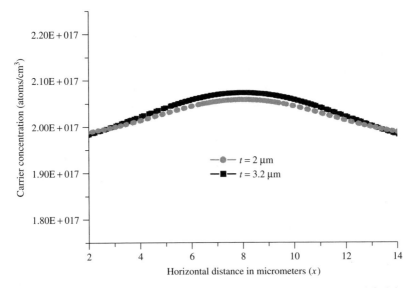

Figure 6.11 Effect of rib height on injected carriers along $y = 5.5$. Modified from Figure 7 in C K Tang, G T Reed, A J Walton, and A G Rickman (1994) 'Low loss, single mode optical phase modulator in SIMOX material', *J. Lightwave Technol.*, **12**, 1397. Reproduced by permission of IEEE

Table 6.1 Effect of rib width on peak carrier concentration. Modified from Table 1 in C K Tang, G T Reed, A J Walton, and A G Rickman (1994) 'Low loss, single mode optical phase modulator in SIMOX material', *J. Lightwave Technol.*, **12**, 1397. Reproduced by permission of IEEE

Rib width (μm)	3	4	5
Peak carrier concentration along $y = 3.2$ ($\times 10^{17}$ cm^{-3})	2.09	2.12	2.06
Peak carrier concentration along $y = 2$ ($\times 10^{17}$ cm^{-3})	2.15	2.17	2.18

change due to the variation in device area), a comparison can be made simply by comparing the peak carrier injection at corresponding points in the devices. In order to evaluate the effect of rib width on peak carrier concentration, three variants of the device were modelled, with rib widths of 3 μm, 4 μm and 5 μm. All other parameters were as shown in Figure 6.7. The results are given in Table 6.1. It is clear that the rib width has little effect on the injected carrier density. Therefore, in an application where rib width is important, it could be varied whilst remaining confident of achieving predictable carrier concentrations.

Depth of the n^+ and p^+ Regions

For the device of Figure 6.7, the effect of the depth of the n^+ and p^+ regions can be evaluated by simulation. However, there would be a detrimental effect if the p^+ region were made any deeper since it would encroach onto the region of the optical mode, causing significant absorption. Therefore in this work only the depth of the n^+ regions was varied. It may be expected that such changes would vary the injected carrier density significantly, but the simulations suggest that the variation is rather modest. In order to consider the variation in depth, four depths were considered: 0.5 μm, 1.2 μm, 1.75 μm and 2.5 μm, a significant variation. The concentration of the n^+ regions was assumed to remain constant at 5×10^{18} cm^{-3}. This is a significant assumption since it implies a fabrication technique other than diffusion, perhaps ion implantation. The results suggested that there may be some optimum depth, because the 1.2 μm depth produced the highest level of injection, although only marginally higher than the other depths evaluated. An example result is shown in Figure 6.12, which clearly shows little variation between devices with different depth contacts. This is particularly interesting because more recent work with different device geometries suggests that a deeper contact significantly improves the device efficiency, particularly if the contact reaches the depth of the buried oxide layer [4]. It should also be borne in mind that the deeper the contact, the more likely the optical mode is to impinge on the doped region, resulting in absorption.

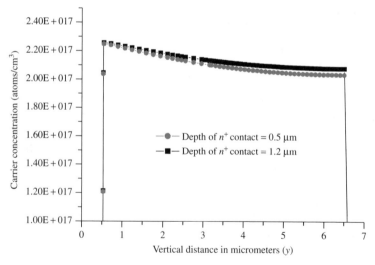

Figure 6.12 Variation in injected carrier density with contact depth. Modified from Figure 8 in C K Tang, G T Reed, A J Walton, and A G Rickman (1994) 'Low loss, single mode optical phase modulator in SIMOX material', *J. Lightwave Technol.*, **12**, 1398. Reproduced by permission of IEEE

Lateral Displacement of n^+ Regions from the Base of the Rib

The lateral translation of the n^+ regions is clearly a design variable. However, the impact of fabrication tolerances may also manifest itself as a translation of the n^+ regions. In order to evaluate the effect of the lateral displacement of the n^+ regions, the device of Figure 6.7 was reproduced for a variety of distances from the base of the rib. As expected, moving the n^+ contacts further apart reduces the injection concentration. This is a direct consequence of increasing the cross-sectional area of the device. An example result comparing distances from the base of the rib to the n^+ region of 0.25 µm, 1 µm and 4 µm is shown in Figure 6.13.

It can be clearly seen from Figure 6.13 that the level of injected carriers is reduced when the contact spacing is increased. Increasing the spacing from the rib base to the n^+ region from 0.25 µm to 1 µm results in a reduction in injected carrier density of approximately 2 % at the centre of the rib, perhaps less than would have been expected. However, increasing the spacing to 4 µm reduces the injection efficiency significantly.

6.1.5   Predicted Device Operation as a Phase Modulator

Based upon the modelling in the preceding sections we can define a device to produce the best performance, based upon the parameters

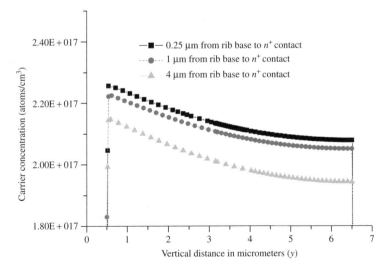

Figure 6.13   Variation in injected carrier density along $x = 8$ for different n^+ contact positions. Modified from Figure 9 in C K Tang, G T Reed, A J Walton, and A G Rickman (1994) 'Low loss, single mode optical phase modulator in SIMOX material', *J. Lightwave Technol.*, **12**, 1399. Reproduced by permission of IEEE

investigated. The device should have sloping rib walls, and a rib height of 3.2 μm (although it would be reasonable to trade this off for simpler fabrication). The n^+ doping contacts should be close to the base of the rib, and a contact depth of the order of 1.2 μm (although once again it would be reasonable to trade this off against ease of fabrication and avoidance of additional absorption).

Such a device delivers an approximately uniform injected carrier density of 2.2×10^{17} cm^{-3}, at a forward bias of 0.9 V. Assuming a wavelength of operation of 1.55 μm, and using equation 4.64, we can evaluate the predicted change in refractive index in the guiding region, Δn, as:

$$
\begin{aligned}
\Delta n = \Delta n_e + \Delta n_h &= -[8.8 \times 10^{-22} \Delta N_e + 8.5 \times 10^{-18} (\Delta N_h)^{0.8}] \\
&= -[8.8 \times 10^{-22} (2.2 \times 10^{17}) + 8.5 \times 10^{-18} (2.2 \times 10^{17})^{0.8}] \\
&= -8.3 \times 10^{-4}
\end{aligned}
\tag{6.1}
$$

The refractive index change results in a phase change of the propagating optical mode, $\Delta \phi$, given approximately by:

$$
\Delta \phi = \frac{2\pi \Delta n L}{\lambda_0}
\tag{6.2}
$$

where L is the length of the active region of the modulator.

A useful benchmark in comparing optical modulators based upon carrier injection is the current required to produce an optical phase shift of π radians for a unit length of the modulator. In this example we can use the concept of a π-radian phase shift to determine the length required to produce such a phase shift, for a fixed level of injection. Therefore using equation 6.2, and setting $\Delta \phi = -\pi$, the required length is:

$$
L_\pi = \frac{\lambda_0}{2\Delta n} = 934 \, \mu m
\tag{6.3}
$$

This is not an unreasonable length for such a device, but is perhaps getting a little long. Consequently the device could be improved by increasing the doping density in the contact regions, which would in turn allow more efficient carrier injection, or simply driven a little harder (i.e. more forward bias), to inject more charge per unit length, and hence obtain more phase change per unit length.

Associated Absorption of the Phase Modulator

We can also predict the additional absorption loss of the same device when forward-biased at 0.9 V. Using equation 4.65, the additional loss,

$\Delta\alpha$, is given by:

$$\Delta\alpha = \Delta\alpha_e + \Delta\alpha_h = 8.5 \times 10^{-18}\Delta N_e + 6.0 \times 10^{-18}\Delta N_h$$
$$= 8.5 \times 10^{-18}(2.2 \times 10^{17}) + 6.0 \times 10^{-18}(2.2 \times 10^{17})$$
$$= 3.19\,\text{cm}^{-1} \tag{6.4}$$

This is equivalent to an enormous 13.9 dB/cm, but since the device is only 934 μm, the additional loss would be 1.3 dB. Because this device discussed in the previous section was a phase modulator, the intention was to minimise the absorption loss (attenuation). However, if we wanted to produce a variable optical attenuator (VOA) we could make the device longer to maximise the attenuation, and also drive the device to a higher potential increasing the level of injected charge. For example if we drive the device harder to produce a level of injection of 5×10^{18}, the attenuation becomes:

$$\Delta\alpha = \Delta\alpha_e + \Delta\alpha_h = 8.5 \times 10^{-18}\Delta N_e + 6.0 \times 10^{-18}\Delta N_h$$
$$= 8.5 \times 10^{-18}(5 \times 10^{18}) + 6.0 \times 10^{-18}(5 \times 10^{18})$$
$$= 72.5\,\text{cm}^{-1} \tag{6.5}$$

This is equivalent to 315 dB/cm, equivalent to 29.4 dB for a device 934 μm in length, or 63 dB for a 2-mm long device. Clearly this is an enormous attenuation, and very useful as a variable optical attenuator, because the attenuation available covers a very large dynamic range.

6.1.6 Fabrication and Experimental Results

A device similar to that described in the previous section was reported by Tang and Reed [2] in 1995. Due to fabrication tolerances, the devices fabricated had different dimensions than those modelled. Therefore, for the purposes of this text it is instructive to consider a device of considerably different dimensions to demonstrate how to compare differing devices. Consider the device shown in Figure 6.14, which was one of the devices fabricated. The device was fabricated with the cross-sectional characteristics shown in the figure, and an active length of 500 μm.

Note that according to the single-mode criteria of Soref et al. [3], this device may not be a single-mode device. However, experimentally only a single mode was observed, suggesting higher-order modes were sufficiently lossy that they were insignificant.

A phase modulator is often fabricated as part of an interferometer, so the modulator causes a phase change in one arm of the interferometer,

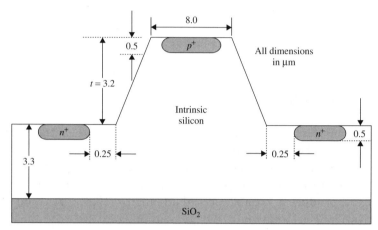

Figure 6.14 Experimental phase modulator. Reproduced from C K Tang and G T Reed (1995) 'Highly efficient optical phase modulator in SOI waveguides', *Electronics Letters*, **31**, 451–452, by permission of IEE

with reference to the static phase in the second arm of the interferometer (for example, a Mach–Zehnder interferometer, the operation of which is discussed later in this chapter). An integrated Mach–Zehnder is shown in Figure 6.15. However, integrating the device with other components introduces the additional losses associated with the other components, and in this case it was desirable to measure the characteristics of the phase modulator in isolation, such as the passive and active loss of the waveguide. Therefore the device was fabricated as part of a straight waveguide. This means that measurement of phase is more difficult, and must be carried out as part of a free-space interferometer. Such a system is shown in Figure 6.16.

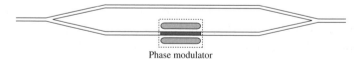

Phase modulator

Figure 6.15 Schematic of a Mach–Zehnder interferometer containing a phase modulator

In the integrated Mach–Zehnder interferometer, interference occurs when the two arms of the interferometer are recombined. The resultant intensity manifests itself as the intensity of the mode in the output waveguide, as discussed later in this chapter. In the free-space interferometer, the two beams from separate paths interfere on a screen (camera in this case), and interference fringes are observed, because the interference is not constrained within the waveguide, and perfect overlay of the two beams is unlikely. Therefore it is a simple matter to measure a π-radian

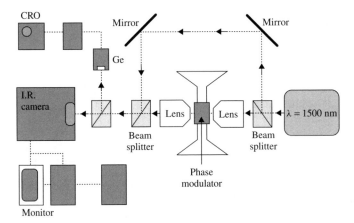

Figure 6.16 Free-space configuration of a Mach–Zehnder interferometer. Reproduced from C K Tang and G T Reed (1995) 'Highly efficient optical phase modulator in SOI waveguides', *Electronics Letters*, **31**, 451–452, by permission of IEE

phase shift in the integrated interferometer, because this corresponds to the shift from maximum intensity to minimum intensity. In the free-space interferometer, it is necessary to monitor the movement of the interference fringes by half of one period. This can be done using video capture techniques.

The experimental devices were measured as described above, and the following results are representative of the devices measured:

- Current required for a π-radian phase shift (I_π): 7 mA
- Passive waveguide propagation loss (modulator off): 0.7 dB/cm
- Additional active optical loss (modulator biased to I_π): 1.3 dB.

Whilst the device has different dimensions than the modelled device, it is possible to make comparisons by considering the current density of each device. The modelled device had a rib width of 4 µm, and a length 934 µm, together with $I_\pi = 4.03$ mA. Since this current flows through the rib surface, the current density, J, at this point is:

$$J_M = \frac{4.03 \times 10^{-3}}{(4 \times 10^{-4})(934 \times 10^{-4})} = 108 \text{ A/cm}^2 \qquad (6.6)$$

Similarly for the experimental device:

$$J_E = \frac{7 \times 10^{-3}}{(8 \times 10^{-4})(500 \times 10^{-4})} = 175 \text{ A/cm}^2 \qquad (6.7)$$

Since the devices are of different lengths, obviously the current densities are different.

Table 6.2 A comparison of theoretical and experimental performance

Parameter	Modelled device	Experimental device
Current density, J	108 A/cm^2	175 A/cm^2
Active optical loss	1.3 dB	1.3 dB
Current required for a π-radian phase shift, I_π	4 mA	7 mA
Refractive index change, Δn	-8.3×10^{-4}	-1.55×10^{-3}
Peak carrier concentration, N	2.2×10^{17} cm^{-3}	4.6×10^{17} cm^{-3}
Figure of merit, χ	17.8°cm/A	20.6°cm/A

It is instructive to compare some other parameters of the two devices. One way to compare devices of different dimensions is to normalise via a figure of merit, χ. In this case, a suitable figure of merit may be defined as:

$$\chi = \frac{\phi}{JL} \qquad (6.8)$$

where ϕ is phase shift in degrees, J is current density, and L is the device length. The figure of merit is included in the comparison Table 6.2.

We can see from Table 6.2 that, despite the differences between the two modulator dimensions, the figure of merit is in reasonable agreement, and is a direct consequence of optimisation via the modelling. This statement has more impact when it is pointed out that, at the time these devices were fabricated, the typical current densities being reported in the literature for carrier injection devices were of the order of kA/cm^2 (e.g. [5]). Therefore this device represented an improvement in the state of the art of approximately an order of magnitude.

6.1.7 Influence of the Thermo-optic Effect on Experimental Devices

Having fabricated a carrier injection device in silicon, the possibility exists that the thermo-optic effect may diminish the efficiency with which the device operates. It is possible to estimate the worst-case effect of the thermo-optic effect very simply. This can be done by adopting a simple one-dimensional model of heat transfer to our device, and assuming that all of the power used to operate the device contributes to heating. In effect this means we are assuming that there is no heat loss from conduction, convection or radiation. This is clearly not the case, but it enables us to calculate the absolute worst-case contribution of the thermo-optic effect. Heat transfer to the device is then given by [2]:

$$P = \frac{\Delta T w L}{(t_{si}/k_{si})} \qquad (6.9)$$

where P is the applied power, ΔT is the temperature rise, w is device width, L is device length, t_{si} is the device thickness (silicon layer), and k_{si} is the thermal conductivity of silicon. For the experimental device reported in the previous section, the applied power (at I_π) was 11.9 mW.

Since the thermal conductivity of silicon at room temperature is $k_{si} = 1.56$ W/cm/K [6], then for an applied power of 11.9 mW the temperature rise ΔT is:

$$\Delta T = \frac{P(t_{si}/k_{si})}{wL} = 0.124\,\text{K} \tag{6.10}$$

We know from equation 4.69 that the corresponding rise in refractive index will be given by:

$$\frac{dn}{dT} = 1.86 \times 10^{-4}/\text{K} \tag{6.11}$$

So:

$$\Delta n = 1.86 \times 10^{-4}(0.124) = 2.3 \times 10^{-5} \tag{6.12}$$

Therefore, even in the worst-case situation where all power contributes to the thermo-optic effect, the change in refractive index due to this effect is very small.

In order to compare the two effects, we can evaluate the ratio of the refractive index changes:

$$\left|\frac{\Delta n_{fc}}{\Delta n_{toe}}\right| = \frac{1.55 \times 10^{-3}}{2.3 \times 10^{-5}} = 67.4 \tag{6.13}$$

That is, the change in refractive index due to the injection of free carriers is 67.4 times larger than that due to the thermo-optic effect. Of course, in reality the ratio will be even larger, as some of the heat will be lost by conduction, convection and radiation. Hence we can conclude that the influence of the thermo-optic effect is negligible in this case. Had the influence not been negligible it would have been worth evaluating the heating effect more accurately (see for example [7]).

6.1.8 Switching Characteristics of the Optical Phase Modulator

The details of the optical phase modulator discussed thus far have all considered the DC performance of the device. However, of paramount importance is also the switching characteristic of such a device, because the phase modulator could form part of an intensity modulator, or indeed an optical switch if configured as part of some form of interferometer

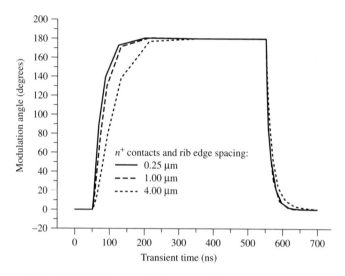

Figure 6.17   Modelled rise and fall times of the optical phase modulator

as discussed above. The switching time then determines the speed of the device, and therefore the range of application for which it is useful.

One of the most convenient ways of determining the switching speed (or time) of the optical phase modulator is to drive it with a fast square wave, and record the rise time of the resulting phase change. We can then characterise the device in terms of its rise and fall times. Figure 6.17 shows the modelled response of the modulator in Figure 6.7, but for the three positions of the n^+ contacts. The three positions are those used for the DC calculations earlier when we were comparing injection efficiency of the carriers (Figure 6.13). Therefore the distances from the base of the rib to the n^+ contact region of 0.25 μm, 1 μm and 4 μm were used for the simulation. For these transient solutions, the device anode and cathode were first zero-biased for 50 ns, followed by a step increase to V_π for 500 ns, and finally a step decrease to 0 V. In each case, V_π is the voltage corresponding to a 180° phase shift. The rise time, t_r, is defined as the time required for the induced phase shift to change from 10 % to 90 % of the maximum value, and the fall time t_f is defined as the time required for the induced phase shift to change from 90 % to 10 % of the maximum value.

We can see from Figure 6.17 that the rise time is significantly slower than the fall time of this device, and is therefore the limiting transition. More importantly we see that the rise time for the device with contacts far from the rib is significantly slower than the device with contacts close to the rib. This is primarily a consequence of the additional time taken for the charges to be moved through a greater distance, and as such is not surprising (although there are also secondary effects that change the rise time). The rise time for the fastest device (0.25 μm spacing) is 58.3 ns, and the fall time is 33.1 ns, corresponding to a bandwidth of the order of 6 MHz. The device with the 1 μm spacing has a rise time of 65.5 ns and a fall time of 31.4 ns, and the device with the 4 μm spacing has a rise time of 117 ns and a fall time of 43.5 ns. This clearly demonstrates the effect on the response time due to changing the device dimensions. Even the fastest of the three devices is a relatively slow device, and certainly not sufficiently fast for optical communications. The device is limited by the fact that switching requires the movement of a significant amount of charge over a fixed distance. Making the device smaller will immediately improve the response time of the devices, because the time taken to move the charge will be reduced. Recent work by Png et al. [8] has shown that silicon devices with bandwidths in excess of 1 GHz are predicted. Given the relatively preliminary nature of this work it is probable that much faster silicon devices will emerge in the future.

6.2   THE MACH–ZEHNDER INTERFEROMETER

Interferometers are central to many optical circuits, but one of the most frequently used interferometers is the famous Mach–Zehnder type. The device is shown schematically in Figure 6.18. The Mach–Zehnder interferometer is a common device in optical circuits, being the basis of several other devices such as modulators, switches and filters. Let us briefly review the operation of the interferometer.

Firstly consider an input wave, at the input waveguide, and let the wave be of TE polarisation. Assuming the waveguide splitter (Y-junction) at the input of the interferometer divides the wave evenly, the intensities in arm 1 and arm 2 of the interferometer will be the same. We can represent the electric fields of the propagating modes in arm 1 and arm 2 of the interferometer as E_1 and E_2 respectively, where:

$$E_1 = E_0 \sin(\omega t - \beta_1 z) \tag{6.14a}$$

$$E_2 = E_0 \sin(\omega t - \beta_2 z) \tag{6.14b}$$

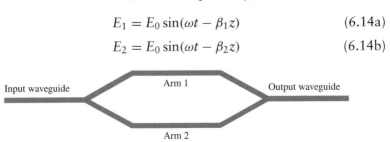

Figure 6.18   Schematic of a waveguide Mach–Zehnder interferometer

For the moment we have considered the two fields to have the same amplitude, but different propagation constants. The two fields propagate along their respective arms of the interferometer and recombine at the output waveguide. When the input Y-junction divides the input field, the two fields formed in arm 1 and arm 2 will be in phase. However, when the fields recombine, they may no longer be in phase, either due to different propagation constants in the arms, or due to different optical path lengths in the arms. The intensity at the output waveguide, S_T, will be:

$$S_T = [(E_1 + E_2) \times (H_1 + H_2)] = S_0(E_1 + E_2)^2 \qquad (6.15)$$

Therefore we need to evaluate the term $(E_1 + E_2)^2$. Let us assume different path lengths for the two waves, of L_1 and L_2. Expanding and substituting for E_1 and E_2 gives:

$$S_T = S_0\{E_0^2 \sin^2(\omega t - \beta_1 L_1) + E_0^2 \sin^2(\omega t - \beta_2 L_2)$$
$$+ 2E_0^2 \sin(\omega t - \beta_1 L_1) \sin(\omega t - \beta_2 L_2)\} \qquad (6.16)$$

Using trigonometric identities, equation 6.16 can be rewritten as:

$$S_T = S_0\{E_0^2(\tfrac{1}{2}[1 - \cos(2\omega t - 2\beta_1 L_1)]) + E_0^2(\tfrac{1}{2}[1 - \cos(2\omega t - 2\beta_2 L_2)])$$
$$+ E_0^2[\cos(\beta_2 L_2 - \beta_1 L_1) - \cos(2\omega t - \beta_2 L_2 - \beta_1 L_1)]\} \qquad (6.17)$$

Since optical frequencies are very high, only the time average of these waves can be observed. Hence all terms in equation 6.17 must be replaced with their time average equivalent, yielding equation 6.18:

$$S_T = S_0 \left\{ \frac{E_0^2}{2} + \frac{E_0^2}{2} + E_0^2[\cos(\beta_2 L_2 - \beta_1 L_1)] \right\}$$
$$= S_0\{E_0^2[1 + \cos(\beta_2 L_2 - \beta_1 L_1)]\} \qquad (6.18)$$

Equation 6.18 is the well-known transfer function of the interferometer, which is plotted in Figure 6.19, normalised to a maximum amplitude of 1. The term $(\beta_2 L_2 - \beta_1 L_1)$ represents the phase difference between the waves from each arm of the interferometer. Clearly if the two arms of the interferometer have identical waveguides, and hence identical propagation constants, the transfer function will have maxima when the path length difference, $|L_2 - L_1|$, results in a phase difference of a multiple of 2π radians. Similarly the transfer function will have mimima when the phase difference is a multiple of π radians.

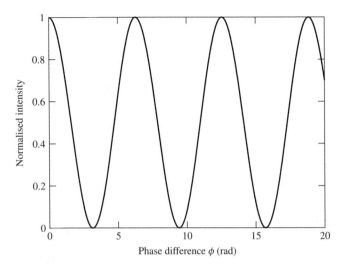

Figure 6.19 Normalised transfer function of a Mach–Zehnder interferometer

Thus the intensity at the output of the Mach–Zehnder interferometer can be manipulated to be a maximum or a minimum, by varying the relative phase of the two arms of the interferometer. This could be accomplished by inserting an optical phase modulator into one of the interferometer arms. In the case of the silicon modulator discussed in the previous section, the unwanted additional absorption due to free carriers will result in an imbalance in the intensities in each of the interferometer arms, and consequently, in imperfect interference in the output waveguide, and in particular in an imperfect 'null' in the transfer function. If such an intensity modulator is to be used over less than 2π radians, one arm can be biased in terms of loss, to compensate.

6.3   THE WAVEGUIDE BEND

The majority of the waveguides discussed so far in this text have been depicted as simple straight structures, uniform in the z direction (except for the waveguides forming the Mach–Zehnder interferometer). However, to make the optical circuits practical we need to be able to send light to various parts of the circuit. This means that we must be able to form bends in the waveguide. Alternatively, one could imagine a series of straight waveguides joined together, but the abrupt junction of such waveguides would result in scattering centres, and hence losses would result at each abrupt change. A curved waveguide allows a gradual transition from one direction to another that can have negligible loss.

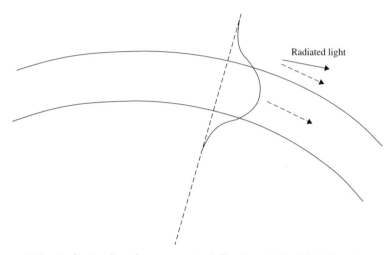

Figure 6.20   Radiation loss from an optical fibre bend. Modified from Figure 3.4 in John M Senior, *Optical Fiber Communications - Principles and Practice*, second edition. Reproduced by permission of Pearson Education

At first sight, it may seem extravagant to regard a simple curve in a waveguide as a separate device, and hence perhaps it is unexpected to find bends covered in this chapter. However, like many other optical devices, the bend requires careful design in order that it is not lossy. For this reason it is included in this chapter, so that we can use the knowledge we have gained in earlier chapters to understand the design issues for the waveguide bend.

Consider Figure 6.20, which shows an optical fibre bend top view. An illustration of the lateral optical field is also shown, including the evanescent fields that extend into the cladding. It is convenient to consider the optical fibre, as the symmetrical nature of the fibre simplifies the explanation of loss. We will consider a rib waveguide subsequently.

Consider the mode shown in Figure 6.20, travelling around the fibre bend. Because the arc of the bend at the outside of the bend is longer than the arc at the inside of the bend, light at the outer cladding must propagate more quickly than light at the inner cladding, in order to maintain the phase relationship across the mode. As the evanescent tail extends into the outer cladding, eventually a distance is reached where light in the outer cladding will need to exceed the velocity of unguided light in the same material, in order to maintain the mode. This is, of course, impossible, so light is radiated and lost from the mode.

A similar situation occurs in a waveguide mode, although the geometrical shape of the waveguide is more complex. Such a situation is shown in Figure 6.21.

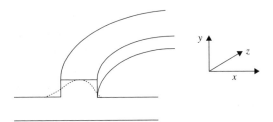

Figure 6.21 Waveguide bend showing distorted lateral mode shape. Adapted with permission from *Introduction to Semiconductor Integrated Optics*, by Hans P Zappe, 1995, Artech House Publishing, Norwood, MA, USA, www.artechhouse.com

Marcatili and Miller [9] carried out an analysis of such a waveguide bend to determine the loss coefficient. Whilst their analysis was not for a rib waveguide, the use of appropriately calculated propagation constants and decay constants inside and outside the rib gives reasonable results for a rib waveguide. They used an assumption that waveguide radiation from a bend was similar to emission of photons from an abruptly ended waveguide. This enabled them to determine a criteria for defining when light was lost from the guide, and hence define a distance from the side of the bend when light could be considered sufficiently far to no longer be part of the propagating mode. This is a consequence of having to consider the distortion of the mode travelling around a bend, as shown in Figure 6.21, and to decide when light is sufficiently far from the waveguide to be considered lost. However, their analysis was based upon modes defined within the straight waveguide. Nonetheless, their result is widely used, partly due to the mathematical convenience. They showed that the loss coefficient from the bend was of the form:

$$\alpha_{\text{bend}} = C_1 \exp(-C_2 R) \tag{6.19}$$

where R is the bend radius, and C_1 and C_2 are related to the waveguide and mode properties. The constants C_1 and C_2 are given by:

$$C_1 = \frac{\lambda_0 \cos^2\left(k_{\text{xg}} \frac{w}{2}\right) \exp(k_{\text{xs}} w)}{w^2 k_{\text{xs}} n_{\text{effp}} \left[\frac{w}{2} + \frac{1}{2k_{\text{xg}}} \sin(w k_{\text{xg}}) + \frac{1}{k_{\text{xs}}} \cos^2\left(k_{\text{xg}} \frac{w}{2}\right)\right]} \tag{6.20}$$

$$C_2 = 2k_{\text{xs}} \left(\frac{\lambda_0 \beta}{2\pi n_{\text{effp}}} - 1\right) \tag{6.21}$$

where β is the z-directed propagation constant, k_{xg} is the x-directed propagation constant in the waveguide, k_{xs} is the x-directed decay constant representing the evanescent field, w is the waveguide width, n_{effp} is the effective index outside the rib, and λ_0 is the free-space wavelength.

Consideration of equation 6.19 shows that the loss coefficient is critically dependent upon the radius of curvature of the bend. Consequently the radius of curvature must be as large as possible to minimise loss. However, for most applications a small device footprint is desirable, implying that the radius should be as small as possible. Furthermore, the constants C_1 and C_2 are critically dependent upon the x-directed loss coefficient, k_{xs}. In both constants, reducing the value of k_{xs} reduces the loss. Hence this tells us that increasing the confinement of the waveguide will reduce the loss from the bend. An alternative way to look at this statement is that tightly confining waveguides can tolerate tighter bends for a given bend loss. Thus the silicon-on-insulator technology is potentially a good candidate for small bend dimensions, as it is a highly confining technology owing to the large refractive index of silicon. Of course the degree of lateral confinement will vary with the etch depth of the rib waveguide, and hence careful design of the rib associated with a bend is very important. To demonstrate the losses involved, as predicted by this model, let us consider both small and large silicon-on-insulator waveguide structures.

Consider the rib waveguide of Figure 6.22. Firstly let us consider a waveguide with the following parameters: $h = 5\,\mu m$, $r = 3\,\mu m$ and $w = 3.5\,\mu m$. For these conditions, and assuming TE polarisation at a wavelength of $\lambda = 1.55\,\mu m$, we can evaluate the bend radius from equation 6.19, for a bend loss of 0.1 dB/cm. The resulting bend radius is approximately 7.3 mm, rather large. It is worth noting that this figure agrees reasonably well with the sort of figures achieved experimentally, giving some confidence in the model. See for example the results of Rickman and Reed [10] for large waveguides, with bend radii of the

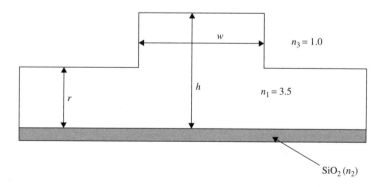

Figure 6.22 Rib waveguide geometry

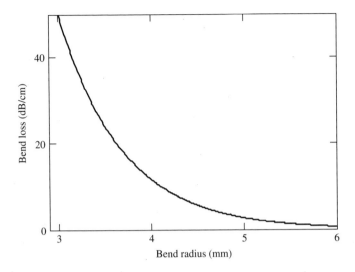

Figure 6.23   Bend loss variation with bend radius for a large rib waveguide

order of 7–10 mm. However, it should also be remembered that this is a simple model of a waveguide bend, and so is rather approximate. It is helpful to consider the variation of bend loss with radius of the bend. This is plotted in Figure 6.23 for the waveguide defined above.

Before moving to a smaller waveguide, it is interesting to consider the effect of etching the waveguide a little more deeply, and hence better confining the mode. Let the conditions now be: $h = 5\,\mu m, r = 2.5\,\mu m$ and $w = 3.5\,\mu m$. Therefore the waveguide sidewalls have been etched by a further half of one micrometre. The radius of a waveguide with a loss of 0.1 dB/cm now falls to only 3.0 mm, significantly smaller.

Let us now consider a much smaller waveguide. Let the waveguide parameters be: $h = 2\,\mu m, r = 1.2\,\mu m$ and $w = 1.4\,\mu m$. Proportionally these are the same dimensions as for the large rib discussed above. However, the bend radius of a waveguide with a loss of 0.1 dB/cm now falls to 0.74 mm. We can plot the variation in loss as a function of bend radius, as we did for a larger rib. This is shown in Figure 6.24.

In a similar fashion to the previous example, let us now etch a little deeper into a small waveguide, so that the parameters become: $h = 2\,\mu m, r = 1.0\,\mu m$ and $w = 1.4\,\mu m$. In this case the waveguide now can tolerate a bend radius of only 310 microns. Clearly the smaller the waveguides become, the smaller the bend radius can be, hence saving on valuable real-estate area on the silicon wafer. This is just one of the reasons why there is currently a trend towards smaller photonic circuit dimensions.

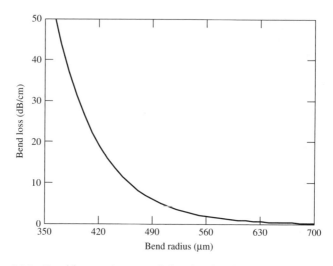

Figure 6.24 Bend loss variation with bend radius for a small rib waveguide

6.4 THE WAVEGUIDE-TO-WAVEGUIDE COUPLER

In Chapters 2 and 3 we saw that power propagating along a waveguide includes some power travelling outside the waveguide in the cladding. This was characterised in our planar waveguide by the penetration depth of the cladding. The field that extends beyond the waveguide is referred to as the *evanescent field*. We can use this field to couple light from one waveguide to another if the waveguides are sufficiently close that the evanescent fields overlap, in a device called a waveguide-to-waveguide coupler, or an evenescent coupler. Zappe [11] presented a useful simplified analysis of such a device, and we follow his approach here.

Consider the two identical waveguides, (a) and (b), shown in Figure 6.25. The waveguides have width w, and are separated by a small distance s. Let the field in waveguide (a) be described by the equation:

$$E_a = a_0(x, y)e^{j\beta z}e^{jwt} \qquad (6.22)$$

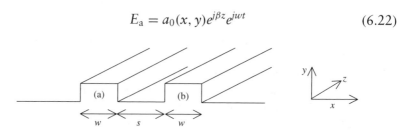

Figure 6.25 Two waveguides separated by a small distance s. Adapted with permission from *Introduction to Semiconductor Integrated Optics*, by Hans P Zappe, 1995, Artech House Publishing, Norwood, MA, USA, www.artechhouse.com

Similarly the field in waveguide (b) will be:

$$E_b = b_0(x, y)e^{j\beta z}e^{jwt} \tag{6.23}$$

The way the fields interact can be described by coupled mode theory (see for example [12]). Simplified versions of the coupling equations that relate the amplitudes within each waveguide are:

$$da_0/dz = \kappa b_0 \tag{6.24}$$

$$db_0/dz = -\kappa a_0 \tag{6.25}$$

where κ is a coupling coefficient. In these simplified equations we have assumed that the waveguides are identical and hence a phase match exists between the modes in each waveguide, no attenuation, and co-directional coupling. If we now assume that one of the waveguides is excited at $z = 0$, and the other is not, then we have:

$$a_0(z = 0) = 0 \tag{6.26}$$

$$b_0(z = 0) = c_0 \tag{6.27}$$

The solutions of these equations take the form:

$$a_0(z) = c_0 \sin(\kappa z) \tag{6.28}$$

$$b_0(z) = c_0 \cos(\kappa z) \tag{6.29}$$

Therefore it is clear that a field in one waveguide gives rise to a field in the other, over some propagation distance z, and further that the transfer of power from one guide to the other is periodic, with a period given by the coupling length, referred to as L_π. We can see that complete transfer of power from one guide to the other takes place when $L_\pi = m\pi/2\kappa$, for integer values of m. Therefore we can also imagine coupling lengths that are a specific fraction of the coupling length to couple a predetermined proportion of the power from one waveguide to the other. A particularly common example is the coupler with coupling length $L_c = L_\pi/2 = m\pi/4\kappa$, which couples half of the energy from one waveguide to the other. This structure is known as the 3-dB coupler for obvious reasons.

The fact that power transfers from one guide to another in a periodic manner means that over multiple coupling lengths the power will transfer back and forth between the waveguides. This means that if a single coupling length is very short, it may be more convenient to make devices with multiple coupling lengths, although care must be taken to ensure that fabrication tolerances are not exacerbated in this case.

The coupling efficiency is determined by the coupling coefficient, κ, which can be expressed as [13]:

$$\kappa = \frac{2k_{xc}^2 k_{xs} \exp(-k_{xs}s)}{\beta w (k_{xs}^2 + k_{xc}^2)} \qquad (6.30)$$

where k_{xc} is the x-directed propagation constant in the core, k_{xs} is the x-directed decay constant (between the waveguides), w is the waveguide width, and s is the waveguide separation. The exponential term in equation 6.30 is clearly a strong influence on the value of κ. This shows that the waveguide spacing and decay constant are very important parameters to the coupler. Clearly this is expected, as the degree to which the evanescent fields overlap is dependent upon these parameters.

Equation 6.30 shows that the coupling coefficient is not only a function of modal confinement, but also of the propagation constant. This implies that coupling efficiency will vary with wavelength, and hence by cascading a series of couplers of slightly different design it is possible to achieve a specific wavelength dependence.

6.4.1 Applications of the Waveguide-to-Waveguide Coupler

The waveguide-to-waveguide coupler is a fundamental device to a number of other devices. Whilst it is beyond the scope of this text to give exhaustive examples, two important examples are briefly discussed. Firstly, consider a Mach–Zehnder interferometer formed with a waveguide-to-waveguide coupler replacing one or both of the waveguide Y-junctions. An example is shown in Figure 6.26.

We have seen from section 6.4 that light will be in one or both of the output waveguides depending on the design of the coupler as well

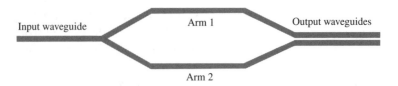

Figure 6.26 Waveguide-to-waveguide coupler at the output of a Mach–Zehnder interferometer

Figure 6.27 Waveguide-to-waveguide coupler forming part of an optical ring resonator

as the interaction length of the coupler. Hence it is possible to design the coupler such that varying the phase of the optical wave results in switching from one output waveguide to the other, forming an optical switch. Furthermore, by also replacing the input Y-junction of the interferometer, a 2 × 2 switch would result.

The waveguide-to-waveguide coupler can also be used as an input and/or output device for other devices. A ring resonator is one such example, shown in Figure 6.27. The input waveguide excites the resonator, by coupling a proportion of the input power to the ring. The device will then act as an interferometer, because after each revolution of the ring, the light in the ring will be in phase with incoming light, only if the phase shift introduced by propagation around the ring, $\Delta\phi$, is an integral number of wavelength periods; i.e.:

$$\Delta\phi = 2m\pi \tag{6.31}$$

$$\beta L = 2m\pi \tag{6.32}$$

where β is the waveguide propagation constant, L is the optical path length, and m is an integer. Thus the device will be resonant for wavelengths that satisfy this condition. Substituting for the optical path length, $L = 2\pi R$, where R is the radius of the ring, and for β in terms of wavelength, we can write an expression for resonant wavelengths of the ring:

$$\lambda = \frac{2\pi NR}{m} \tag{6.33}$$

where N is the effective index of the waveguide mode. An additional coupler can be added at the opposite side of the ring resonator to obtain an output in antiphase to the first, since only half the phase shift is experience by one half revolution of the ring.

6.5   THE ARRAYED WAVEGUIDE GRATING (AWG)

The AWG is one of the most important integrated optical devices introduced in recent years. The device has proved to be very flexible, being utilised in a number of configurations including multiplexing of wavelengths, demultiplexing, switching and $N \times N$ routing, and as part of add–drop filter designs. This section is included to introduce the AWG, and provide a basic understanding of the operation of the device. Firstly however, it is helpful to consider the interference of a series of coherent light sources, as this will aid the understanding of the AWG.

6.5.1   Interference of N Coherent Light Sources

Whilst the interference of several coherent sources will not fully describe the behaviour of a diffraction grating, or the AWG, it is a useful starting point, as we can understand the ideas behind the AWG. This is a well-known interference problem, and in this text we will follow the approach of Hecht [14].

Consider a linear array of N identical oscillators, a distance d apart. Initially consider the oscillators to be both coherent and in phase, and of the same polarisation. This situation is shown schematically in Figure 6.28. Consider interference of the rays from all of the sources (S_1 to S_N) at some distant point, P. If the size of the array is small compared to the distance to P, the amplitude arriving at point P will be approximately the same from each source. Furthermore the rays from the sources will be approximately parallel arriving at the distant point P.

We can express each contribution to the electric field arriving at point P as:

$$E_i = E_0 \exp[j(\omega t - \beta r_i)] \qquad (6.34)$$

where the subscript i represents all sources from 1 to N, and r is the distance from each source to the point P. Therefore the total field arriving at P, E_t, is the sum of the individual contributions:

$$E_t = E_0 \exp[j(\omega t - \beta r_1)] + E_0 \exp[j(\omega t - \beta r_2)]$$
$$+ \cdots + E_0 \exp[j(\omega t - \beta r_N)] \qquad (6.35)$$

It can be seen from Figure 6.28 that the additional path length, Δr, to P between successive sources is $d \sin \theta$. Therefore the additional phase shift, $\Delta \phi$, at P from successive sources is:

$$\Delta \phi = -\beta \Delta r = -\beta d \sin \theta \qquad (6.36)$$

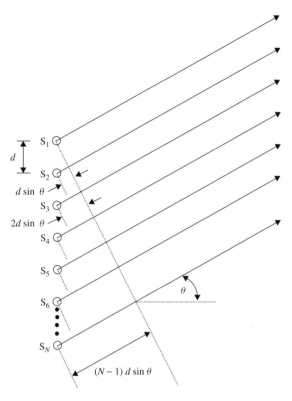

Figure 6.28 N coherent sources in a linear array. Source: Figure 10.7, p440 from OPTICS, 3rd ed. by Eugene Hecht. Copyright © 1998 by Addison Wesley Longman, Inc. Reprinted by permission of Addison Wesley Longman Publishers, Inc.

Therefore the total field at P may be rewritten as:

$$E_t = E_0 \exp[j(\omega t - \beta r_1)] \times [1 + \exp(j(\Delta\phi)) + \exp(j(2\Delta\phi))$$
$$+ \cdots + \exp(j(N\Delta\phi))] \tag{6.37}$$

However, the term in the second square bracket is just a geometric series, so equation 6.37 reduces to:

$$E_t = E_0 \exp[j(\omega t - \beta r_1)] \left[\frac{\exp(j(N\Delta\phi)) - 1}{\exp(j(\Delta\phi)) - 1} \right]$$
$$= E_0 \exp[j(\omega t - \beta r_1)] \left[\frac{\exp(jN\Delta\phi/2)}{\exp(j\Delta\phi/2)} \right]$$
$$\times \left[\frac{\exp(jN\Delta\phi/2) - \exp(-jN\Delta\phi/2)}{\exp(j\Delta\phi/2) - \exp(-j\Delta\phi/2)} \right]$$

$$= E_0 \exp[j(\omega t - \beta r_1)][\exp(j(N-1)\Delta\phi/2)]$$

$$\times \left[\frac{\exp(jN\Delta\phi/2) - \exp(-jN\Delta\phi/2)}{\exp(j\Delta\phi/2) - \exp(-j\Delta\phi/2)} \right]$$

$$= E_0 \exp[j(\omega t - \beta r_1 + (N-1)\Delta\phi/2)] \left[\frac{\sin(N\Delta\phi/2)}{\sin(\Delta\phi/2)} \right] \quad (6.38)$$

If we now consider the intensity at point P, which is proportional to $(E_t)^2$, we obtain:

$$I = I_0 \left[\frac{\sin^2(N\Delta\phi/2)}{\sin^2(\Delta\phi/2)} \right] \quad (6.39)$$

where I_0 is the intensity from any single source arriving at P. We can also substitute for $\Delta\phi$ in order to see the dependence on θ. Recalling from equation 6.36 that:

$$\Delta\phi = -\beta\Delta r = -\beta d \sin\theta \quad (6.40)$$

then equation 6.39 becomes:

$$I = I_0 \left[\frac{\sin^2[(N\beta d/2)\sin\theta]}{\sin^2[(\beta d/2)\sin\theta]} \right] \quad (6.41)$$

We can plot equation 6.41 for various values of N (the number of sources). This is shown in Figure 6.29, for 2, 5 and 10 sources. Clearly

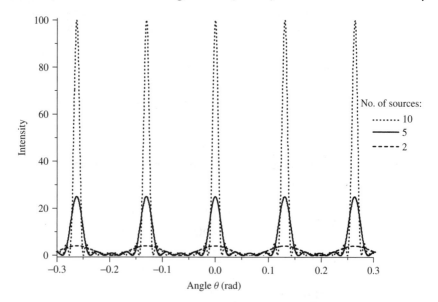

Figure 6.29   Intensity variations with number of sources N

the principal peaks of the curve occur when $\Delta\phi = 2m\pi$, which is a function of angle θ. These principal peaks increase in intensity with the square of the number of sources, as expected, because when $\Delta\phi = 2m\pi$, equation 6.41 reduces to $I = N^2 I_0$. Notice that there are also subsidiary maxima between the principal peaks, owing to the rapid variations of the numerator in equation 6.41.

Let us now introduce an additional phase shift between adjacent sources. If we make the additional phase shift equivalent to a fixed change in path length, ΔL, then the phase shift will vary with wavelength of operation. This is because the phase shift is given by the propagation constant multiplied by the path length. For each wavelength, the propagation constant will be different, and hence the phase shift will be different.

Consider the effect on the interference pattern of an array of 10 sources, as shown in Figure 6.30. The first significant difference is that the principal maxima are no longer spaced symmetrically about an angle of $\theta = 0$. This means that introducing a successive phase shift to all sources can shift the position of the principal maxima. Secondly, when we introduce slightly different wavelengths, because the phase shift is slightly different in each case, the principal peaks occur at different positions. In the case of no additional phase shift the principal

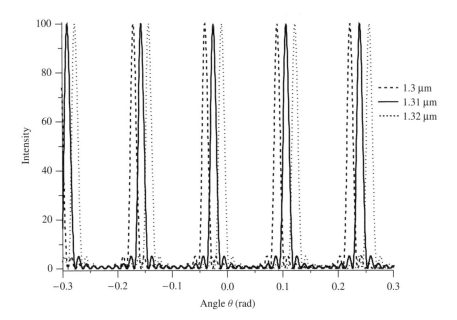

Figure 6.30 Addition of a fixed path length difference can separate wavelengths

peaks occurred when:

$$|\Delta\phi| = \beta d \sin\theta = 2m\pi \qquad (6.42)$$

When additional path length ΔL is included, the principal peaks now occur when:

$$|\Delta\phi| = \beta d \sin\theta + \beta\Delta L = 2m\pi \qquad (6.43)$$

Now that we have demonstrated that an increasing phase change across an array of coherent sources can result in separation of wavelengths, let us consider the operation of the arrayed waveguide grating itself.

6.5.2 Operation of the AWG

The purpose of this section of the text is to give an explanation of the operation of the AWG. The most powerful operation of the AWG is the separation of wavelengths within an integrated optical device, so by way of example the operation of an AWG as a wavelength demuliplexer will be described. This section follows the approach of [15], to explain the outline operation of the AWG. The general layout is shown in Figure 6.31.

It can be seen that the AWG comprises two planar regions which act as passive star couplers, and an array of rib waveguides, of progressively increasing length. The first planar region is used to excite the array of

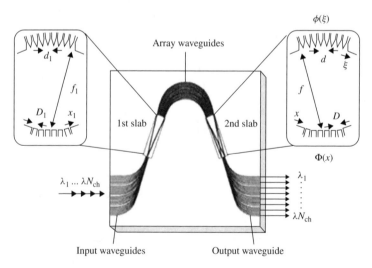

Figure 6.31 Configuration of a typical AWG layout. Reprinted from Academic Press, San Diego, CA, K Okamoto, *Fundamentals of Optical Waveguides*, © 2000, with permission from Elsevier

waveguides, and the second planar region is to allow multiple beam interference from the outputs of the array of waveguides. The array waveguides are included to introduce incremental phase shifts to the rays emerging from those waveguides. If a single wavelength is introduced into one of the input waveguides, this wavelength is distributed to the array waveguides. This wavelength propagates through these waveguides, each ray emerging with the incremental phase shift due to the length of the waveguide in question. This is equivalent to introducing an incremental phase shift to the series of N coherent sources in section 6.5.1. In the second planar region the beams emerging from the array waveguides interfere to produce a pattern with a single principal peak, that spatially coincides with one of the output waveguides. If a second wavelength is introduced into the same input waveguide, it will also be distributed to the array of waveguides, but will experience a different phase shift through each of the waveguides, owing to the different propagation constant associated with each wavelength. Consequently the peak of the interference pattern of the second wavelength in the second planar region will occur at a different output waveguide, separating the two wavelengths. Multiple wavelengths can be separated in this way, producing a wavelength demultiplexer.

We can consider the demultiplexer operation in more detail. Let the array of rib waveguides have a constant path length difference, ΔL, between neighbouring waveguides. Following the notation of [15], in the first slab region, let the input waveguide separation be D_1, the array waveguide separation be d_1, and the radius of curvature be f_1. In general, the waveguide parameters in the first and the second slab regions do not have to be the same, therefore in the second slab region, let the output waveguide separation be D, the array waveguide separation be d, and the radius of curvature be f. Measuring the input position anticlockwise from the centre of the input waveguides as position x_1, the light is radiated to the first slab and then excites the arrayed waveguides. The excited electric field amplitude at each array waveguide is a_i ($i = 1$ to N), where N is the total number of array waveguides. The amplitude profile of a_i is typically a gaussian distribution (although more complex distributions can be more favourable). After travelling through the arrayed waveguides the light beams constructively interfere into one point x, measured in an anticlockwise direction from the centre of the output waveguides in the second slab, in the same way multiple beam interference was demonstrated in section 6.5.1. Therefore, the position of this focal point depends on the operating wavelength and the relative phase delay in each waveguide, which is given by $\Delta L / \lambda$.

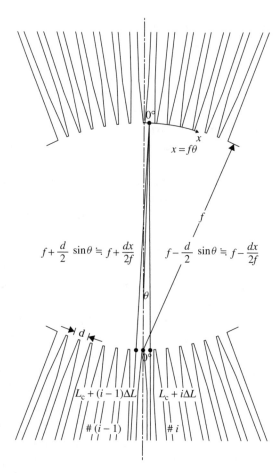

Figure 6.32 Enlarged view of the output optical slab of the AWG. Reprinted from Academic Press, Diego, CA, K Okamoto San, *Fundamentals of Optical Waveguides*, © 2000, with permission from Elsevier

Figure 6.32 shows an enlarged view of the second slab. If we consider the total phase delays for the two light rays passing through the $(i-1)$th and ith array waveguides, the geometrical distances of the two beams in the second slab region are approximated as shown in Figure 6.32. The first slab will be of similar configuration, but may have different dimensions. The difference of the total phase delays for the two light rays passing through the $(i-1)$th and ith array waveguides must be an integral multiple of 2π in order that the two beams constructively interfere at the focal point x. Since the phase change is given by the propagation constant multiplied by the distance of propagation, if we assume identical propagation constants in each of the two slabs, then the interference condition is:

$$\beta_s \left(f_1 - \frac{d_1 x_1}{2f_1} \right) + \beta_c (L_c + (i-1)\Delta L) + \beta_s \left(f + \frac{dx}{2f} \right)$$

$$= \beta_s \left(f_1 + \frac{d_1 x_1}{2f_1} \right) + \beta_c (L_c + i\Delta L) + \beta_s \left(f - \frac{dx}{2f} \right) - 2m\pi \quad (6.44)$$

where β_s and β_c denote the propagation constants in the slab region and array waveguide, m is an integer, λ_0 is the centre wavelength of the WDM system, and L_c is the minimum array waveguide length. Eliminating common terms we obtain:

$$\beta_s \frac{d_1 x_1}{f_1} - \beta_s \frac{dx}{f} + \beta_c \Delta L = 2m\pi \quad (6.45)$$

When the phase shift is a multiple of 2π, $\beta_c \Delta L = 2m\pi$. So:

$$\lambda_0 = \frac{N_c}{m} \Delta L \quad (6.46)$$

Then the light input position x_1 and the output position x will be related by:

$$\frac{d_1 x_1}{f_1} = \frac{dx}{f} \quad (6.47)$$

In equation 6.46, N_c is the effective index of the array waveguides, and m is the *diffraction order*. Therefore, when light is coupled into the input position x_1 the output position x is determined by equation 6.47. Often the waveguide parameters in the first and second slab regions are the same, and then the input and output distances are equal ($x_1 = x$). The variation of the focal position x with respect to wavelength λ, for the fixed input light position x_1, can be found by differentiating 6.44 with respect to λ as [15]:

$$\frac{\Delta x}{\Delta \lambda} = -\frac{N_g f \Delta L}{N_s d \lambda_0} \quad (6.48)$$

where N_s is the effective index in the slab region, and N_g is the group index related to the effective index N_c of the array waveguides, defined as $N_g = N_c - \lambda dN_c / d\lambda$. Similarly, the variation of the input side position

x_1 with respect to wavelength λ for the fixed light out position x is given by:

$$\frac{\Delta x_1}{\Delta \lambda} = \frac{N_g f_1 \Delta L}{N_s d_1 \lambda_0} \tag{6.49}$$

The input and output waveguide spacings are $|\Delta x_1| = D_1$ and $|\Delta x| = D$ respectively when $\Delta \lambda$ is the channel spacing. By inserting these relations into 6.48 and 6.49, the wavelength spacing in the output slab can be found for a fixed light input position x_1:

$$\Delta \lambda_{out} = \frac{N_s d D \lambda_0}{N_g f \Delta L} \tag{6.50}$$

and similarly the wavelength spacing at the input, for the fixed light output position x, is:

$$\Delta \lambda_{in} = \frac{N_s d_1 D_1 \lambda_0}{N_g f_1 \Delta L} \tag{6.51}$$

When the waveguide parameters in the first and second slab regions are the same, then the channel spacings are the same ($= \Delta \lambda$). The path length difference, ΔL, can be found from 6.50 as:

$$\Delta L = \frac{N_s d D \lambda_0}{N_g f \Delta \lambda} \tag{6.52}$$

The spatial separation of the mth and $(m + 1)$th focused beams for the same wavelength is given from 6.43 as:

$$X_{FSR} = x_m - x_{m-1} = \lambda_0 f / N_s d \tag{6.53}$$

X_{FSR} is called the free spatial range of the AWG, in the same way that different diffraction orders of other gratings and interferometers are used to specify the free spectral range of those devices.

The number of wavelength channels, N_{ch}, that can be utilised is found by dividing X_{FSR} by the output waveguide separation D as:

$$N_{ch} = \frac{X_{FSR}}{D} = \frac{\lambda_0 f}{N_s d D} \tag{6.54}$$

The theory of the optimum shape of the arrayed waveguides will not be discussed here, but there are a number of restrictive conditions for several waveguide parameters. The minimum bend radius should be obviously be larger than the known minimum bending radius of the rib waveguides used in the design, as discussed earlier in this chapter. The straight lengths should be larger than the length required to taper the waveguide array elements into the slab in a lossless manner, and the minimum array waveguide separation at the centre should be larger than the minimum coupling width so that the ribs do not interact in the manner of a waveguide-to-waveguide coupler. Although there are many design possibilities for a given AWG specification, the best design is the one in which the array waveguide lengths are as short as possible so as to minimise phase errors.

The frequency response of a typical 40-channel AWG is shown in Figure 6.33. The crosstalk is less than 30 dB. Devices have been manufactured with up to 128 ports with channel spacing of 12.5 GHz.

An AWG provides a fixed routing of an optical signal from a given input port to a given output port based on the wavelength of the signal. Signals at different wavelengths coming into an input port will each be routed to a different output port. Also, different signals using the same wavelength can be input simultaneously to different input ports and still not interfere with each other at the output ports. Compared to a passive star coupler in which the given wavelength may only be used on a single input port, the AWG with N input and N output ports is capable of routing a maximum of N^2 connections, as opposed to a maximum of N connections in the passive star coupler. This is shown schematically in Figure 6.34.

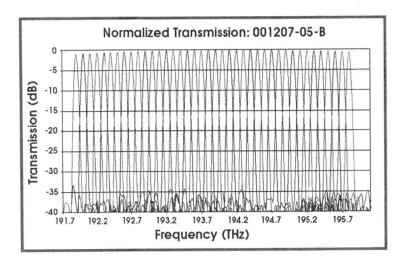

Figure 6.33   40-channel AWG frequency response. Reproduced by permission of Bookham Technology PLC

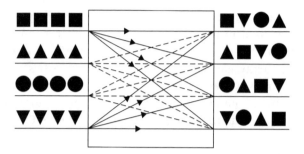

Figure 6.34 Arrayed waveguide grating router (functional schematic). Reproduced from G P Agrawal (1997) *Fiber Optic Communications Systems*, 2nd edn, by permission of John Wiley & Sons, New York

However, if there are N input and N output ports and n possible wavelengths on each input port, because of the cyclic nature of the distribution of the wavelengths between the input ports and output ports, a fixed routing structure is obtained. This means that different orders of diffraction are effectively being used at the same time in the AWG. This causes difficulties because the outer ports suffer higher losses than the central ports. The main disadvantage of the AWG is that it is a device with a fixed routing matrix, so it cannot be reconfigured.

6.6   WAVEGUIDE COUPLERS FOR SMALL-DIMENSION WAVEGUIDES

In Chapter 4 we noted the trend to smaller waveguide dimensions, and the associated difficulty of coupling to small waveguides. In this section we will study two possible solutions to this problem in the form of two distinct devices. These are the *three-dimensional taper* and the *dual grating-assisted directional coupler* (DGADC).

Let us first consider the three-dimensional taper. Examples of such work has been published by Confluent Photonics [17], and by Prather et al. [18], as shown in Figure 6.35. Authors typically quote losses of between 0.7 dB and 1 dB per interface for three-dimensional tapers, although this figure rises sharply for cross-sectional dimensions of the waveguide smaller than approximately 2 μm.

An alternative technique based upon grating was introduced by Butler et al. [19], who proposed a grating-assisted directional coupler. A schematic of a similar type of device is shown in Figure 6.36. The idea

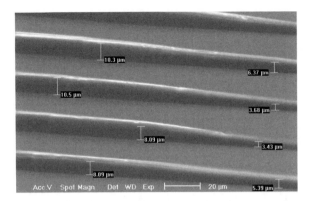

Figure 6.35 An example of vertical tapers in silicon, to act as couplers to small waveguides. Source: D W Prather, J Murakowski, S Venkataraman et al. (2003) 'Novel fabrication methods for 2D photonic crystals in silicon slab waveguides', *Proc. SPIE*, **4984**, (2003). Reproduced by permission of SPIE

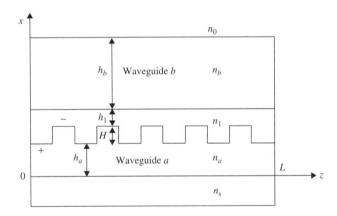

Figure 6.36 Grating-assisted directional coupler. Modified from Figure 1 in G Z Masanovic, V M N Passaro, and G T Reed (2003) 'Coupling optical fibres with thin semiconductor waveguides', *Proceedings of the SPIE*, **4997**, 172. Reproduced by permission of SPIE

is that light can be efficiently coupled from an optical fibre to the large surface waveguide, waveguide *b*, and then the grating is used to couple light to the thin silicon waveguide, waveguide *a*. Unfortunately, waveguides *a* and *b* need to be so different in both dimensions and refractive index that only low-efficiency coupling results. In the work of Butler et al. [19], the maximum coupling efficiency, for TE polarisation, was only 40 % for optimised waveguide and grating parameters. Furthermore, for a change in the grating period of just 0.3 nm, the coupling efficiency dropped by almost 50 %, making the fabrication of

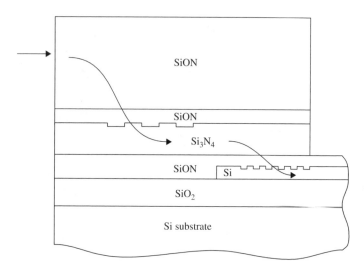

Figure 6.37 Dual grating-assisted directional coupler in SOI. Source: G Z Masanovic, V M N Passaro and G T Reed 'Dual grating-assisted directional coupling between fibres and thin semiconductor waveguides', *Photonics Technol. Lett.*, 15, 1395–1397 © 2003 IEEE

this grating-assisted directional coupler extremely difficult to realise, and impractical for commercial applications.

In 2003, a new approach based upon grating was proposed by Masanovic et al. [20,21], the dual grating-assisted directional coupler, which consists of two conventional gratings linked by one layer with the value of refractive index that lies between the fibre refractive index and semiconductor refractive index. This layer is of crucial importance for achieving high coupling efficiency. The device is shown in Figure 6.37. The DGAGC is similar to the coupler in Figure 6.35 in that light is coupled from a fibre into a large surface waveguide, and then transferred to a thin silicon waveguide. However, the use of two gratings and an intermediate layer gives sufficient additional flexibility to optimise the coupling process, and to achieve grating dimensions that are relatively straightforward to fabricate. The authors demonstrated a theoretical coupling efficiency to a waveguide of only 250 nm in thickness, of up to 96 %, corresponding to a coupling loss of less than 0.18 dB.

However, at the time of writing this text, no satisfactory coupler was experimentally available for coupling with high efficiency to waveguide of less than about 2 μm in cross-sectional dimensions, although those devices discussed above show significant potential. This is clearly a challenge for silicon photonics in the next few years.

REFERENCES

1. C K Tang, G T Reed, A J Walton and A G Rickman (1994) 'Low loss single mode optical phase modulator in SIMOX material', *IEEE J. Lightwave Technol.*, **12**, 1934–1400.

2. C K Tang and G T Reed (1995) 'Highly efficient optical phase modulator in SOI waveguides', *Electron. Lett.*, **31**, 451–452.

3. R A Soref, J Schmidtchen and K Petermann (1991) 'Large single-mode rib waveguides in GeSi–Si and Si-on-SiO$_2$', *J. Quantum Electronics*, **27**, 1971–1974.

4. P D Hewitt and G T Reed (2000) 'Improving the response of optical phase modulators in SOI by computer simulation', *IEEE J. Lightwave Technol.*, **18**, 443–450.

5. J P Lorenzo and R A Soref (1987) '1.3 μm electro-optic switch', *Appl. Phys. Lett.*, **51**, 6–8.

6. M N Wybourne (1988) 'Thermal conductivity of Si', in *Properties of Silicon*, Emis Data Review Series no. 4, INSPEC (IEE), London.

7. B A Moller, L Jensen, C Laurent-Lund and C Thirstrup (1993) 'Silica waveguide thermo-optic phase shifter with low power consumption and low lateral heat diffusion', *IEEE Photon. Tech. Lett.*, **5**, 1415–1418.

8. C E Png, G T Reed, R M H Atta, G Ensell and A G R Evans (2003) 'Development of small silicon modulators in silicon-on-insulator (SOI)', *Proc. SPIE*, **4997**, 190–197.

9. E A J Marcatili and S E Miller (1969) 'Improved relationships describing directional control in electromagnetic wave guidance', *Bell Syst. Tech. J.*, **48**, 2161–2188.

10. A G Rickman and G T Reed (1994) 'Silicon on insulator optical rib waveguides: loss, mode characteristics, bends and y-junctions', *IEE Proc. Optoelectron.*, **141**, 391–393.

11. H A Zappe (1995) *Introduction to Semiconductor Integrated Optics*, Artech House, London, ch. 8.

12. R G Hunsperger (1991) *Integrated Optics: Theory and Technology*, 3rd edn, Springer-Verlag, Berlin.

13. S Somekh, E Garmire, A Yariv, H L Garvin and R G Hunsperger (1973) 'Channel optical waveguide directional couplers', *Appl. Phys. Lett.*, **22**, 46–47.

14. E Hecht (1998) *Optics*, 3rd edn, Addison-Wesley, London, ch. 10.

15. K Okamoto (2000) *Fundamentals of Optical Waveguides*, Academic Press, San Diego, CA.

16. G P Agrawal (1997) *Fiber Optic Communications Systems*, 2nd edn, John Wiley & Sons, New York.

17. 'Coupling of single mode fibres to planar Si waveguides using vertically tapered mode converters', www.confluentphotonics.com.

18. D W Prather, J Murakowski, S Venkataraman et al. (2003) 'Novel fabrication methods for 2D photonic crystals in silicon slab waveguides', *Proc. SPIE*, **4984**, 89–99.
19. J K Butler, N H Sun, G A Evans, L Pang and P Congdon (1998) 'Grating-assisted coupling of light between semiconductor and glass waveguides', *J. Lightwave Technol.*, **16**, 1038–1048.
20. G Z Masanovic, V M N Passaro and G T Reed (2003) 'Coupling optical fibres with thin semiconductor waveguides', *Proc. SPIE*, **4997**, 171–180.
21. G Z Masanovic, V M N Passaro and G T Reed (2003) 'Dual grating-assisted directional coupling between fibres and thin semiconductor waveguides', *Photonics Technol. Lett.* **15**, 1395–1397, 2003.

7
Polarisation-dependent Losses: Issues for Consideration

The response of the optical circuit to the incoming optical signal should ideally be the same regardless of the polarisation of that optical beam. However, in practice there are a variety of reasons why the response is different. This is important because the polarisation of the incoming signal may vary, particularly if it originates from a circularly symmetrical optical fibre, which typically provides a signal of random polarisation. Hence there is not even the opportunity to compensate for a known degree of polarisation imbalance, because the relative contributions of each polarisation to the total signal may change with time. Consequently designers usually aim to minimise any polarisation dependence in integrated optical devices.

The net result of different responses to the polarisation of the incoming signal usually manifests itself as a different signal loss to each polarisation, although the origins of that loss will not usually be limited to an inherent differential loss, but may also be derived from differential phase shifts, different degrees of optical confinement, or differential optical paths lengths within the circuit. The difference in loss between the orthogonal polarisation components of the signal is usually termed *polarisation-dependent loss* (PDL), and it is the issues related to PDL that form the material of this chapter. Some of the primary contributions to PDL will be discussed in turn, being pulled together in the discussion and conclusion sections.

7.1 THE EFFECT OF WAVEGUIDE THICKNESS

Let us consider the silicon planar waveguide shown in Figure 7.1. We observed in Chapter 2 that the eigenvalue equations for TE and TM

Silicon Photonics: An Introduction   Graham T. Reed and Andrew P. Knights
© 2004 John Wiley & Sons, Ltd   ISBN: 0-470-87034-6

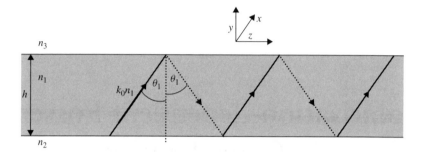

Figure 7.1 Propagation in a planar waveguide

polarisations are different. This means that the modal solutions to these equations will also be different. For completeness the equations are reproduced here. The eigenvalue equation for TE modes is:

$$[k_0 n_1 h \cos\theta_1 - m\pi] = \tan^{-1}\left[\frac{\sqrt{\sin^2\theta_1 - (n_2/n_1)^2}}{\cos\theta_1}\right]$$

$$+ \tan^{-1}\left[\frac{\sqrt{\sin^2\theta_1 - (n_3/n_1)^2}}{\cos\theta_1}\right] \quad (7.1)$$

Similarly the eigenvalue equation for TM modes is:

$$[k_0 n_1 h \cos\theta_1 - m\pi] = \tan^{-1}\left[\frac{\sqrt{(n_1/n_2)^2 \sin^2\theta_1 - 1}}{(n_2/n_1)\cos\theta_1}\right]$$

$$+ \tan^{-1}\left[\frac{\sqrt{(n_1/n_3)^2 \sin^2\theta_1 - 1}}{(n_3/n_1)\cos\theta_1}\right] \quad (7.2)$$

To demonstrate the differences between polarisation modes, let us consider a series of planar silicon waveguides, and plot the intensity mode profiles of the fundamental mode for both TE and TM polarisations. There is relatively little difference between the mode profiles of asymmetric and symmetric waveguides in silicon, so it is convenient to use the symmetric waveguide, particularly since in many practical applications the surface of the silicon wafer will usually have a passivating layer of SiO$_2$, making the waveguide a symmetrical one. In this case the eigenvalue equations for the TE and TM modes (equations 7.1 and 7.2 above) reduce to:

For TE polarisation:

$$\tan\left[\frac{k_0 n_1 h \cos\theta_1 - m\pi}{2}\right] = \left[\frac{\sqrt{\sin^2\theta_1 - (n_2/n_1)^2}}{\cos\theta_1}\right] \quad (7.3)$$

For TM polarisation:

$$\tan\left[\frac{k_0 n_1 h \cos\theta_1 - m\pi}{2}\right] = \left[\frac{\sqrt{(n_1/n_2)^2 \sin^2\theta_1 - 1}}{(n_2/n_1)\cos\theta_1}\right] \quad (7.4)$$

If we now determine the fundamental intensity mode profiles for both polarisations, of four symmetrical planar waveguides, we can immediately see the effect of polarisation. Consider waveguides with the following parameters: $n_1 = 3.5$ (silicon), $n_2 = n_3 = 1.5$ (silicon dioxide), $\lambda_0 = 1.3\,\mu$m. If we now determine the fundamental intensity mode profiles for waveguides with heights of $0.3\,\mu$m, $1.0\,\mu$m, $5\,\mu$m and $8\,\mu$m, we would expect the confinement of the fundamental modes to increase with increasing height h. The results are shown in Figures 7.2, 7.3, 7.4 and 7.5.

It is clear that in general the confinement of the TE and TM modes is different. This means that if the waveguide cladding is lossy in some way, or the core/cladding interface is lossy, then the modes will experience different losses because there is more power in the cladding (and at the interface) for one mode than the other. It is also clear that as the waveguide is increased in size, the TE and TM modes become similar, and the polarisation dependence is reduced. It is also worth noting in passing that the mode profiles presented above are normalised to have unity peak intensity. This has been done so that a direct comparison can be made. It is more usual to normalise these profiles to unity power flow, but in general this results in different amplitudes when plotting both profiles on the same axis, and hence visual comparison is more difficult.

The differences in confinement of the modes can be understood by considering again the graphical solution of the eigenvalue equation, as demonstrated in Chapter 2. If we plot both the TE and the TM eigenvalue equations on the same axes (Figure 7.6), it is clear that the

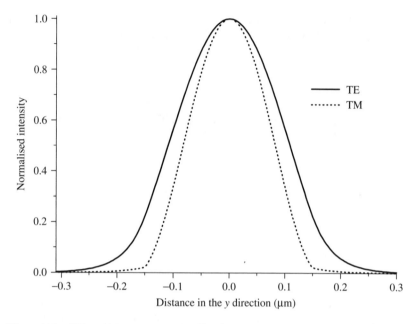

Figure 7.2   TE and TM intensity profiles for a waveguide height of $h = 0.3\,\mu\text{m}$

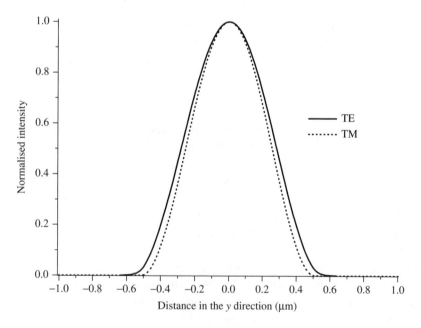

Figure 7.3   TE and TM intensity profiles for a waveguide height of $h = 1\,\mu\text{m}$

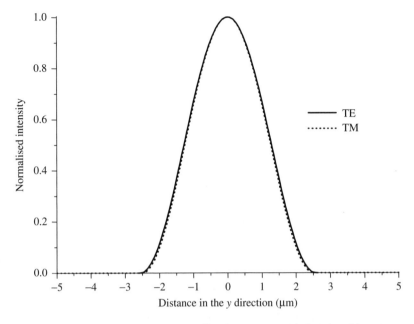

Figure 7.4 TE and TM intensity profiles for a waveguide height of $h = 5\,\mu\text{m}$

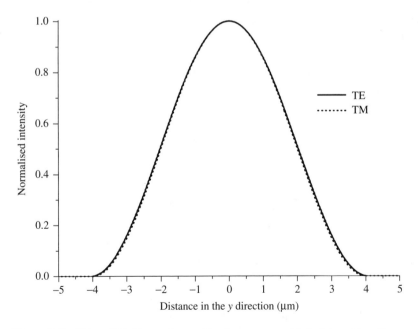

Figure 7.5 TE and TM intensity profiles for a waveguide height of $h = 8\,\mu\text{m}$

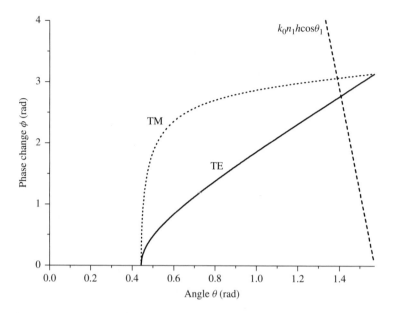

Figure 7.6   Solution of the TE and TM eigenvalue equation for $h = 1\,\mu\text{m}$

phase change on reflection for the TM mode is always greater than that for the TE mode. Figure 7.6 shows these graphs for the waveguide parameters associated with Figure 7.3, when $h = 1\,\mu\text{m}$.

Since the intersection of the curve of $k_0 n_1 h \cos\theta_1$ and the TE curve is always to the right of the intersection with the TM curve, the TE mode always has a larger propagation angle than the TM mode. Recall that the lateral (y-directed) propagation constant was given by the following equation:

$$k_y = n_1 k_0 \cos\theta_1 \qquad (7.5)$$

Since θ_1 is larger for the TE mode, then $\cos\theta_1$ is smaller for the TE mode, and hence the propagation constant, k_y, is also smaller. Therefore the phase change across the waveguide is also smaller for the TE mode, resulting in the field profiles of Figures 7.2 to 7.5.

Note that Figure 7.5, for the largest planar waveguide, suggests that the mode profiles are coincident. This is just a function of the scale of the diagram. If we look more closely at the region of one of the core/cladding boundaries, we still see the separation of the modes. For example, taking the $8\,\mu\text{m}$ waveguide of Figure 7.5 and 'zooming in' at the upper boundary results in Figure 7.7.

It is clear, therefore, that increasing the waveguide thickness brings the TE and TM modes closer together in terms of their field profiles.

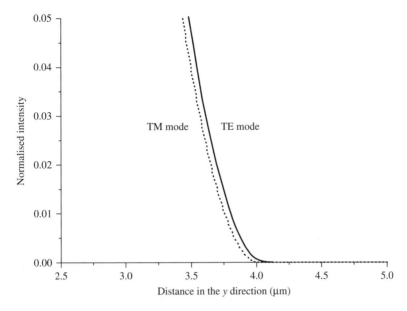

Figure 7.7   Enlarged view of part of the TE and TM mode profiles for $h = 8\,\mu m$

However, there is still a finite separation between the degree of confinement, although in practice the difference may be negligible for some applications. The situation is more complex for rib waveguides, and will be discussed further in Section 7.4.

7.2   SURFACE SCATTERING LOSS FOR DIFFERENT WAVEGUIDE THICKNESS AND POLARISATION

The previous section demonstrated that TE and TM modes experience different degrees of optical confinement, even in a waveguide fabricated in a material that is nominally isotropic. One of the consequences of this is that the power at the waveguide core/cladding interfaces will be different for the two polarisations. Also, because the propagation angles are different, one mode will experience more reflections at the core/cladding interface than the other. In turn, this means that the loss due to scattering at the interfaces will also be different, because this loss is related to the intensity at the interfaces, together with the degree of interaction with the interfaces. Recall that in Chapter 4 an approximate model of interface scattering was introduced, originally produced by Tien in 1971 [1], and based upon the specular reflection of power from a surface. By considering the total power flow over a given distance, together with the loss at both waveguide interfaces, Tien produced the following expression for the loss coefficient due to interface scattering:

$$\alpha_s = \frac{\cos^3 \theta}{2 \sin \theta} \left(\frac{4 \pi n_1 (\sigma_u^2 + \sigma_\ell^2)^{\frac{1}{2}}}{\lambda_0} \right)^2 \left(\frac{1}{h + \frac{1}{k_{yu}} + \frac{1}{k_{y\ell}}} \right) \quad (7.6)$$

where σ_u is the r.m.s. roughness for the upper waveguide interface, σ_ℓ is the r.m.s. roughness for the lower waveguide interface, k_{yu} is the y-directed decay constant in the upper cladding (as defined in Chapter 2), $k_{y\ell}$ is the y-directed decay constant in the lower cladding, and h is the waveguide thickness.

In order to further investigate the polarisation-dependent loss of the waveguides of section 7.1, we can evaluate the scattering loss of the fundamental TE and TM modes of each of these waveguides in turn. In order to do this, we must first evaluate the decay constants in the cladding for each waveguide. Since the waveguides are symmetrical, the decay constant for the upper cladding, k_{yu}, and the decay constant for the lower cladding, $k_{y\ell}$, will be equal. Evaluating these decay constants in turn for waveguides of thickness 0.3 μm, 1.0 μm, 5 μm and 8 μm, and for each polarisation, gives the results in Table 7.1.

Following Chapter 4, if we now let both σ_u and σ_ℓ equal 1 nm, we can evaluate the scattering loss for each fundamental mode, for each nanometre of r.m.s. roughness at the interfaces. Using equation 7.6 we obtain the results in Table 7.2.

Notice that despite the TM mode being more tightly confined, according to Tien's model, it experiences a higher loss. This is rather suspicious, and so it is instructive to compare these results with another scattering model, of which there are many. If we follow the approach of, for example, Marcuse [2], Lacey and Payne [3], and Ladouceur et al. [4],

Table 7.1 Decay constants for a series of planar waveguides of increasing thickness

Waveguide thickness (μm)	TE$_0$ $k_{yu} = \sqrt{\beta^2 - k_0^2 n_2^2}$	TM$_0$ $k_{yu} = \sqrt{\beta^2 - k_0^2 n_2^2}$
0.3	13.481 μm^{-1}	11.968 μm^{-1}
1	15.0297 μm^{-1}	14.97322 μm^{-1}
5	15.2717 μm^{-1}	15.2712 μm^{-1}
8	15.2791 μm^{-1}	15.27896 μm^{-1}

Table 7.2 Interface scattering loss for planar waveguides of increasing thickness, using Tien's model

Waveguide thickness (μm)	Interface scattering loss TE$_0$ (dB/cm)	Interface scattering loss TM$_0$ (dB/cm)
0.3	9.453	22.832
1	0.1966	0.2656
5	4.6×10^{-4}	4.9×10^{-4}
8	7.29×10^{-5}	7.58×10^{-5}

these authors developed models based upon a correlation function over the length of the waveguide. Using an exponential correlation function, a simple model for the loss coefficient results as follows [3,4]:

$$\alpha_s = \Phi^2 \left(\frac{h}{2} \right) (n_{core}^2 - n_{cladd}^2)^2 \left(\frac{k_0^2}{8n_{core}} \right) \left(\frac{1}{N - n_{cladd}} \right) \sigma^2 \qquad (7.7)$$

where n_{core} is the refractive index of the core, n_{cladd} is the refractive index of the cladding, N is the effective index of the propagating mode, h is the waveguide thickness, and $\Phi^2(h/2)$ is the value of $[E_x(y)]^2$ in the TE case, or $(1/n^2)[H_x(y)]^2$ in the TM case, at the core/cladding interface. In both cases power normalisation is used such that $\int_{-\infty}^{\infty} \Phi^2(y)dy = 1$.

As would be expected, there are several similarities between equations 7.6 and 7.7, although only the latter explicitly uses the power at the core/cladding interface to evaluate the loss coefficient. If we repeat the calculations of loss coefficient for the waveguide parameters of Table 7.2, using the model of equation 7.7, the results in Table 7.3 are obtained.

Clearly the results from the latter model are different from those of Tien's model. This merely highlights the limitations of various models, and the need to take care in applying the results of specific models to

Table 7.3 Interface scattering loss for planar waveguides of increasing thickness, using the second model

Waveguide thickness (μm)	Interface scattering loss TE$_0$ (dB/cm)	Interface scattering loss TM$_0$ (dB/cm)
0.3	2.15	3.8
1	0.108	0.063
5	1.13×10^{-3}	5×10^{-4}
8	2.85×10^{-4}	1.23×10^{-4}

a given situation. In particular it shows that the desire to use a simple analytical model, whilst being attractive, is often inherently approximate.

Comparing the results of Tables 7.2 and 7.3, we see that the agreement is worse when the waveguide is small. This is not surprising as many models become inaccurate at dimensions of the order of the wavelength. In the present case the latter model is likely to be more accurate as it has been developed in conjunction with real roughness measurements, and is based upon field intensities at the core/cladding interface. It also predicts lower losses that more closely resemble the results in the literature. In particular, a significant difference between the models is that Tien's model predicts a higher loss from the better-confined TM mode, whereas the model of Ladouceur et al. predicts the opposite.

The results so far are based upon an r.m.s. interface roughness of 1 nm at the core/cladding interface. By extension we can consider the loss due to a range of values of interface roughness. Consider, for example, one of the waveguides above, with a thickness of 1μm. Let us consider the interface scattering loss for r.m.s. roughnesses of 0.1 nm, 0.5 nm, 1 nm, 10 nm and 50 nm, using the model of Ladouceur et al. The results are shown in Table 7.4.

It is also instructive to carry out the same calculation for a larger waveguide, say the 5μm waveguide. The results are shown in Table 7.5.

Table 7.4 Interface scattering loss for a 1μm planar waveguide (fundamental mode)

R.m.s. interface roughness (nm)	Interface scattering loss TE_0 (dB/cm)	Interface scattering loss TM_0 (dB/cm)
0.1	1.1×10^{-3}	6.3×10^{-4}
0.5	0.027	0.016
1	0.108	0.063
10	10.81	6.30
50	270	157

Table 7.5 Interface scattering loss for a 5μm planar waveguide (fundamental mode)

R.m.s. interface roughness (nm)	Interface scattering loss TE_0 (dB/cm)	Interface scattering loss TM_0 (dB/cm)
0.1	1.13×10^{-5}	5×10^{-6}
0.5	2.83×10^{-4}	1.25×10^{-4}
1	1.13×10^{-3}	5×10^{-4}
10	0.113	0.05
50	2.83	1.25

From Tables 7.4 and 7.5 it can clearly be seen that the degree of loss is related not only to the polarisation, but also to the degree of wall roughness, as a function of the waveguide thickness. Consequently if a given fabrication process results in an absolute waveguide roughness, the loss is minimised in larger waveguides, as is the polarisation-dependent loss.

It is clear from the foregoing results that the polarisation-dependent loss can become important if the interface quality is not kept under control, or if the device is very long. For example the interface roughness of the silicon/buried oxide layer of commercially available SIMOX wafers is typically in the range 0.8–3 nm. Even if we assume the surface of such wafers is perfectly smooth, the model of Ladouceur et al. suggests a loss ranging from 0.04 dB/cm to 0.57 dB/cm for the fundamental TM mode of a 1 μm waveguide.

To maintain simplicity, the preceding discussion has focused on planar waveguides, but of course it is also important to consider the scattering loss of three-dimensional waveguides such as rib waveguides. In a rib waveguide, the contribution to loss can be increased as compared to a planar waveguide, because the additional etched silicon surfaces are also potential scattering centres due to interface roughness. This situation is compounded by the fact that the etch is an additional fabrication step, that must be constantly monitored to ensure that quality is maintained. The design of the rib waveguide is also important because the vertical and horizontal surfaces of the rib waveguide may exhibit polarisation-dependent loss. Consequently, if one surface is likely to be significantly rougher than the other owing to a particular fabrication step, the waveguide may exhibit unacceptable polarisation-dependent loss. Of course, as the rib waveguide is reduced in dimensions, like the planar waveguide, it will exhibit proportionately more loss, because the mode confinement will reduce. This makes consideration of the waveguide roughness even more important.

7.3   POLARISATION-DEPENDENT COUPLING LOSS

Not only will the mode profiles and the scattering loss of the waveguide vary with polarisation, but perhaps the most fundamental part of utilising an optical circuit will also be affected by polarisation. This is the loss associated with coupling light to the waveguide itself.

The most usual way of coupling to an optical circuit is via an optical fibre. This process was discussed in Chapter 4, where it was noted that the efficiency with which the light is coupled into the waveguide is a function of (i) how well the fields of the excitation and the waveguide modes match; (ii) the degree of reflection from the waveguide facet; (iii) the quality of the waveguide endface; and (iv) the spatial misalignment of the excitation and waveguide fields. There can also be a numerical aperture mismatch in which the input angles of the optical waveguide are not well matched to the range of excitation angles, but this latter term is neglected here. Of these parameters, (iii) and (iv) will be the easiest to optimise, although note that they will not be independent of polarisation. However, we can assume that in a high-quality device the endface is well polished and, for the purposes of this discussion, that perfect alignment is achieved. Similarly, if we assume normal incidence of the exciting optical field, the reflection from the endface will be nominally the same for both polarisations, as shown in Chapter 4. Consequently, the dominant mechanism affecting the coupling efficiency with regard to polarisation is item (i), the overlap between the excitation and modal fields.

In order to consider the effect of polarisation upon the overlap between the two fields when coupling to a rib waveguide, it is convenient to use a commercially available numerical waveguide simulation package, because an analytical solution becomes difficult owing to the complex mode profile of the rib waveguide. In this section of the text, such a simulator has been used to produce a series of curves determining the coupling loss from an optical fibre to a range of rib waveguides, based upon a fixed silicon layer thickness. Note that Fresnel reflection from the waveguide endface is not included in this calculation.

This calculation is more complex than it appears at first sight, because for any given rib width, etch depth and rib height, the optimum alignment will change with the changing mode shape. That is, the optimum vertical position for good coupling will in general be different for every waveguide. Consequently to find the optimum coupling to any waveguide, it is necessary to scan the input mode position vertically across the input facet. If we make the assumption that optimum coupling is achieved for all waveguides, we can then plot optimum coupling for a given rib waveguide height and etch depth, with the variation in rib width.

For each waveguide height, we can therefore plot a pair of curves (TE and TM) for each etch depth, resulting in a whole family of curves when different etch depths are included. Figure 7.8 shows such a family of curves for a silicon overlayer thickness (rib waveguide height) of $h = 5\,\mu m$, and an input field being the fundamental mode of a silica optical fibre with a core diameter of $5\,\mu m$. The coupling efficiency is evaluated for each input polarisation by monitoring the power in the simulated optical field within the waveguide, after a propagation length of $500\,\mu m$, as a proportion of the launch power. In effect this is

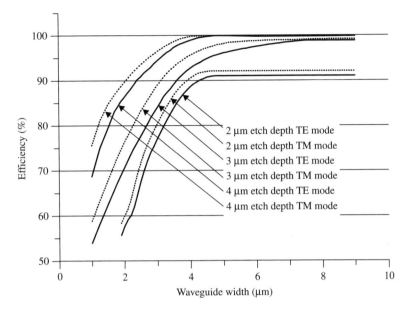

Efficiency (%)

Waveguide width (μm)

2 μm etch depth TE mode
2 μm etch depth TM mode
3 μm etch depth TE mode
3 μm etch depth TM mode
4 μm etch depth TE mode
4 μm etch depth TM mode

Figure 7.8 A family of curves evaluating coupling efficiency from an optical fibre (5-μm core) to various rib waveguides with height $h = 5\,\mu m$

(approximately) an evaluation of the overlap integral of the modes of the waveguide and the fibre.

Individual data points have been evaluated, and trend lines added to guide the eye. It can be seen that the coupling efficiency in most cases is high. This is because the mode sizes of the waveguide and fibre are similar. In general the more confined TM mode couples more efficiently than the TE, but when the waveguide width reaches the same order of dimension as the waveguide height the differences are relatively small. If we carry out a similar evaluation for less well matched modes, less efficient coupling results. This is demonstrated in Figure 7.9. In this case the launch fibre has a core diameter of $9\,\mu m$, and the rib waveguide characteristics are unchanged. In Figure 7.9 the trends are similar, but efficiency continues rising as the waveguide width increases, as the mode widths of fibre and waveguide become similar. Once again the TM mode coupling is slightly more efficient.

Clearly the loss is greater when coupling to the waveguide that is less well matched to the fibre mode. There is also a trend for less polarisation dependence in wider waveguides, and for waveguides with shallower etch depth. However, care should be taken in extracting too much data from these graphs, as coupling efficiency will be related to the mode shapes of all excited modes, and their relative degree

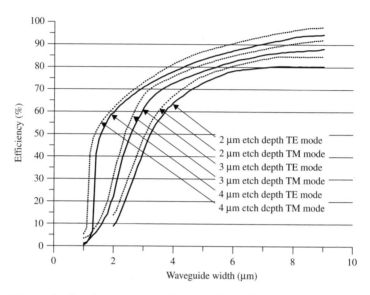

Figure 7.9   A family of curves evaluating coupling efficiency from an optical fibre (9-μm core) to various rib waveguides with height $h = 5\,\mu m$

of excitation, all of which will change with changing rib geometry. Consequently in some cases high-order modes will be well excited and subsequently lost, resulting in lower net coupling efficiency, and in some cases the efficiency of coupling only to the fundamental mode will be high. Consequently detailed modelling should be carried out for each specific case to be considered.

7.4   BIREFRINGENCE

The term 'waveguide birefringence' has emerged from the fibre and integrated optics fields. It is used to describe the difference between either propagation constants or effective indices of the two polarisation modes of the waveguides in question. The term is used indiscriminately to all waveguides regardless of the materials from which they are fabricated, or the origin of the difference in effective indices.

According to the definition of 'birefringence' this is an inaccurate use of the term, because birefringence strictly means the inherent 'double refraction' of light due to the crystal structure of a given material. Double refraction means that if unpolarised light is incident upon a birefringent crystal, two refracted beams rather than one emerge. If the angles of incidence and refraction are considered, only one of the refracted beams obeys Snell's law, as described in Chapters 1 and 2. This is termed the 'ordinary ray' or 'o' ray. The second ray is termed the 'extraordinary ray', or the 'e' ray. The degree of refraction of the extraordinary ray is determined by the degree of birefringence of the crystal in question. Thus crystals that exhibit double refraction are anisotropic, and all crystals except those belonging to the cubic system are anisotropic to some extent. For a more extensive discussion of double refraction, see for example Born and Wolf [5].

Silicon, however, is a cubic crystal, and hence is isotropic, and as such is not inherently birefringent. In practice this means that the refractive index of silicon is constant regardless of the direction of propagation within the crystal. Of course this is not true of an anisotropic crystal such as lithium niobate, also used extensively for integrated optical circuits.

In lithium niobate the ordinary and extraordinary refractive indices are $n_o = 2.29$ and $n_e = 2.21$ [6]. If light is propagating along a principal crystal axis of lithium niobate, the refractive index will be either n_o or n_e, depending on the polarisation of the light, unless the light is propagating along the optical axis, in which case both the e-ray and the o-ray will see a refractive index of n_o. If, however, light is propagating at an arbitrary angle to the principal axes, the ordinary ray will see a refractive index of n_o, but the extraordinary ray will see a refractive index somewhere between n_o and n_e which is determined by the direction within the crystal (see for example [5]). Thus the apparent birefringence within a lithium niobate waveguide depends not only on the geometry of the waveguide itself, but also on the direction of the waveguide within the crystal structure.

This situation is complicated further in lithium niobate by the fact that it is a crystal that exhibits a large electro-optic effect. The refractive index change with applied electric field (as with all electro-optic crystals) is also dependent on the orientation of the field with respect to the crystal. Hence, the crystal orientation is important not only to the birefringence of the crystal, but also to the operation of devices such as electro-optic modulators. Of course it is the very fact that the crystal structure is not cubic that results in the electro-optic effect being present, so the two effects are obviously related. The converse is true for silicon in which the linear electro-optic effect is absent owing to the symmetrical nature of the crystal structure.

7.4.1 Birefringence in Planar Silicon Waveguides

Because silicon is optically isotropic, the apparent waveguide birefringence is due solely to the different propagation constants resulting from the solution of slightly different eigenvalue equations (ignoring parasitic effects such as stress, to be discussed later). Whilst this makes silicon less complex than anisotropic materials, this apparent birefringence can still have serious consequences for devices based on planar waveguides in which relative phase between modes is important.

To consider this further let us evaluate the propagation constants for a planar waveguide of the form discussed in section 7.1. Let the waveguide thickness be 1 μm, and the exciting wavelength be 1.3 μm. If we solve the eigenvalue equations (7.3 and 7.4) for such a waveguide, it is clear that the waveguide will support five TE modes and five TM modes. The propagation constants of each of these modes is listed in Table 7.6.

Recall from Chapter 2 that the phase shift with propagation distance can be evaluated as the product of the propagation constant, β, and the physical length, L. Without specifying a device, let us assume a propagation length of 1 mm, a typical length for some devices and waveguides. The βL product is also evaluated in Table 7.6.

We can immediately see from Table 7.6 that the phase change over 1 mm of each of the modes varies considerably, both with mode number and polarisation. Consequently, any device that relies on particular phase relationships over the length of the device will be seriously compromised by a change from single- to multi-mode behaviour. Similarly, any device which relies on negligible waveguide birefringence over such a propagation distance will be compromised. Consider a simple example in which the device is designed to be a fixed length, but fabrication tolerances result in a device $\delta L = 0.5$ μm longer than designed. The additional phase shift difference between the fundamental TE and first-order TE modes due to the extra length will be:

$$\delta\phi = (\beta_{TE0} - \beta_{TE1})\delta L = (16.687 - 15.983) \times 0.5$$

$$= 0.352 \text{ radians} \equiv 20.2° \qquad (7.8a)$$

Table 7.6 Propagation constants in a 1 μm planar silicon waveguide

Mode number, m	Propagation constant (TE modes)	Propagation constant (TM modes)	Phase change, βL	
			TE	TM
0	$16.687\,\mu m^{-1}$	$16.636\,\mu m^{-1}$	5311.6π	5295.4π
1	$15.983\,\mu m^{-1}$	$15.769\,\mu m^{-1}$	5087.5π	5019.4π
2	$14.751\,\mu m^{-1}$	$14.227\,\mu m^{-1}$	4695.4π	4528.6π
3	$12.880\,\mu m^{-1}$	$11.822\,\mu m^{-1}$	4099.8π	3763.4π
4	$10.150\,\mu m^{-1}$	$8.506\,\mu m^{-1}$	3230.8π	2707.5π

Similarly the phase difference between fundamental TE and TM modes can be evaluated as:

$$\delta\phi = (\beta_{TE0} - \beta_{TM0})\delta L = (16.687 - 16.636) \times 0.5 = 0.0255 \text{ radians}$$
$$\equiv 1.46° \qquad (7.8b)$$

Of course this means that the TE and TM modes will experience a phase difference of 1.46° for every 0.5 μm of waveguide length.

7.4.2 Birefringence in Silicon Rib Waveguides

In common with the planar waveguides, the waveguide birefringence of rib waveguides can be studied by comparing the propagation constants of the modes of the rib waveguides. This is equivalent to studying the effective index difference of modes of the rib waveguides since, from Chapter 4, we know that:

$$\beta = k_0 N_{wg} \qquad (7.9)$$

where N_{wg} is the effective index of the propagating mode.

In order to compare the effect of birefringence on a waveguide device, let us consider the effect of birefringence on a Mach–Zehnder interferometer. A schematic of the interferometer is shown in Figure 7.10.

The polarisation dependence of the Mach–Zehnder will manifest itself in the difference in propagation constants of the waveguides, and the corresponding difference in phase delay in the interferometer. In order to demonstrate this effect in more detail, let us consider a specific example.

Example of a Mach–Zehnder Interferometer

Let the rib waveguides of the Mach–Zehnder have the dimensions shown in Figure 7.11. We can then evaluate the effective indices of the fundamental TE and TM modes. The values of the effective indices are

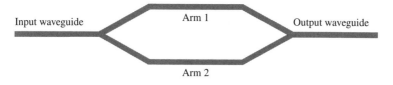

Figure 7.10   Schematic of a waveguide Mach–Zehnder interferometer

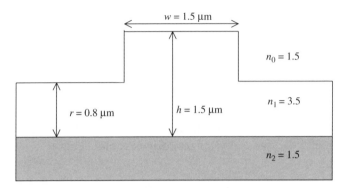

Figure 7.11 Rib waveguide parameters used in the Mach–Zehnder interferometer example

$N_{\text{wg}}(\text{TE}) = 3.4651$ and $N_{\text{wg}}(\text{TM}) = 3.4607$. This corresponds to propagation constants of $\beta(\text{TE}) = 16.748\,\mu\text{m}^{-1}$ and $\beta(\text{TM}) = 16.726\,\mu\text{m}^{-1}$.

If we assume a wavelength of operation of $\lambda = 1.3\,\mu\text{m}$, we can plot the transfer function of the interferometer with increasing path length difference, ΔL, between the two arms of the interferometer. When the path length difference is small, the polarisation dependence will be less noticable, because the phase shift, $\beta\Delta L$, for the TE and TM modes will be similar if ΔL is small. However, as the path length difference is increased, then the phase difference at the output between the two arms for the TE and TM modes will also increase. For the propagation constants evaluated in the example above, the transfer function of the interferometer shows significant polarisation dependence when ΔL reaches a few tens of microns. To demonstrate this, Figure 7.12 shows the two transfer functions when the path length difference is of the order of $100\,\mu\text{m}$.

If the interferometer is designed to be nominally at a null for the TE mode (at $\Delta L = 100.36\,\mu\text{m}$), if a TM mode is also present, the null will be seriously compromised by the TM mode being almost 75 % of its peak value. Of course the relative intensity of the TE/TM interference patterns will also be affected by the amount of optical power in each of the modes, but an equal split would not be an unreasonable assumption if the device is fed by a standard communications fibre, providing randomly polarised light.

We can take this straightforward analysis one stage further by considering what would happen in a waveguide modulator based around a Mach–Zehnder interferometer. We saw in Chapters 4 and 6 the way in which injection of free carriers affects the refractive index of silicon, and can be used to fabricate an optical modulator. If we consider an operating

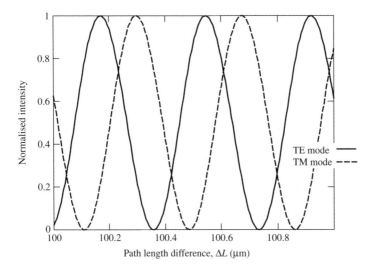

Figure 7.12 Interferometer transfer function for TE and TM modes

wavelength of $1.3\,\mu m$, the refractive index change with injected carriers was described in Chapter 4, reproduced below as equation 7.10:

$$\Delta n = \Delta n_e + \Delta n_h = -[6.2 \times 10^{-22}\Delta N_e + 6.0 \times 10^{-18}(\Delta N_h)^{0.8}] \tag{7.10}$$

If we assume carrier injection levels of $\Delta N_e = \Delta N_h = 1 \times 10^{18}\,\text{cm}^{-3}$, then:

$$\Delta n = -2.13 \times 10^{-3} \tag{7.11}$$

By applying this change in refractive index to the rib waveguides of the Mach–Zehnder above we can consider the relative change in propagation constant of the fundamental mode for TE and TM polarisations. The TE mode propagation constant changes from $\beta(\text{TE}) = 16.748\,\mu m^{-1}$ to $\beta(\text{TE}) = 16.737\,\mu m^{-1}$. Similarly the TM mode propagation constant changes from $\beta(\text{TM}) = 16.726\,\mu m^{-1}$ to $\beta(\text{TM}) = 16.716\,\mu m^{-1}$. Thus, whilst there is nominally no inherent polarisation dependence of carrier injection, there is a small secondary effect that may be significant over very long devices. For example, a carrier injection modulator in silicon may typically be $1000\,\mu m$ in length. The difference in phase shift, $\Delta\phi$, over this length between the modulated and the unmodulated propagation constants for the TE mode is approximately $\Delta\phi(\text{TE}) = 11°$. For the TM mode, approximately $\Delta\phi(\text{TM}) = 10°$. Clearly this small difference is insignificant when compared to the absolute polarisation dependence, over this sort of device length; but if the propagating signals were to pass through a series of cascaded Mach–Zehnder interferometers (in an add–drop filter for example), the effect could become much more significant.

Examples of Birefringence in Rib Waveguides

So far in the discussion of rib waveguides the degree of birefringence for different waveguides has not been explicitly investigated. Following the argument associated with the effective index method, we can think of the vertical and horizontal confinement of an optical mode to be subject to opposing polarisation constraints. That is to say, the polarisation of the nominally TE mode in a rib is subject to 'TE-type' confinement in the vertical sense, and 'TM-like' confinement in the horizontal sense. The converse is true for the TM modes. Therefore we can expect the propagation constants of the TE and TM modes to be affected differently to changes in waveguide geometry such as a change in rib width etc.

Consequently, it is informative to consider the degree of waveguide birefringence for a given rib waveguide height, and how it varies with waveguide width or etch depth. However, before we produce results as an example, let us first consider the model we use to carry out such a task. So far the effective index model has served us well. However, this model can be inaccurate in circumstances associated with high index difference, or dimensions of the order of a wavelength. This becomes particularly important when considering modes with similar propagation constants (and hence similar effective index). This is the case when considering similar order modes of different polarisations, so care must be taken when using the effective index model. To demonstrate this let us evaluate the effective index difference for the fundamental modes of a waveguide, one of TE polarisation, and the other TM.

Consider a large waveguide, with a waveguide height of (say) $5\,\mu m$. If we set the vertical parameter r (see Figure 7.11) to be 3, 2 or $1\,\mu m$ (etch depths of 2, 3 or $4\,\mu m$), we can look at the variation in effective index with varying waveguide width and etch depth. This is plotted in Figure 7.13, using the effective index method to produce the graph. The results are useful in general terms. For example, it can be seen that at larger widths, the etch depth has little effect on the modal birefringence. This is because the 'lateral mode' is well confined for large widths. At small widths the etch depth has more effect, and the birefringence can be reduced. Notice, however, that this may be at the expense of single-mode behaviour, as the geometry also determines whether the waveguide satisfies the single-mode conditions discussed in Chapter 4.

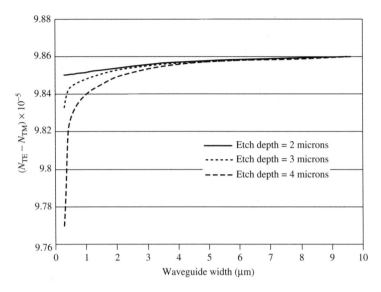

Figure 7.13   Variation of fundamental-mode effective index difference of a $5\,\mu$m waveguide (height) for various etch depths and waveguide widths (effective index method)

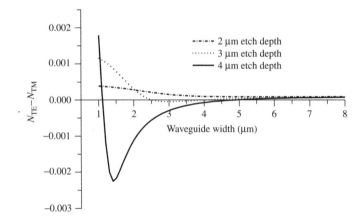

Figure 7.14   Variation of fundamental-mode effective index difference of a $5\,\mu$m waveguide (height) for various etch depths and waveguide widths (beam propagation method)

If we now produce the same graph using a commercial simulator, using the beam propagation method in the simulator, with semi-vectorial capability, Figure 7.14 is the result. There is a striking difference between the graphs of Figures 7.13 and 7.14. Not only are the graph shapes different, but in Figure 7.14 two of the curves cross the 'zero birefringence' axis (i.e. $N_{TE} - N_{TM} = 0$). This indicates that using deeper etch depths makes it is possible to produce birefringence-free waveguides for some of the waveguide geometries and dimensions of Figure 7.14. Figure 7.14 also reflects what we expect to occur in a practical device. For very large waveguide widths, the device will be 'slab-like', and hence the TE effective index will be higher. As the width is reduced the 'TM-like' confinement will compete with 'TE-like' confinement in an 'intermediate stage'. However, when the waveguide width becomes very small, most of the power will be confined to the underlying slab region, and the mode will again exhibit slab-like behaviour and hence higher TE effective indices than TM. The question of whether the curve actually crosses the 'zero birefringence' axis is a function of the relative values of the TE and TM effective indices in the intermediate stage, which is clearly affected by the etch depth.

Now let us consider similar curves for a much smaller waveguide, with say a waveguide height of only 1 μm. Evaluating the difference in effective index for the fundamental modes in this case results in Figure 7.15. In this case two etch depths are considered, of $e = 0.5$ and 0.75 μm. We see similar trends in both the smaller and larger waveguides, but notice that for the smaller waveguides of Figure 7.15, the zero birefringence axis is crossed only for the deeper of our two etch depths, which is a substantial proportion of the overall waveguide

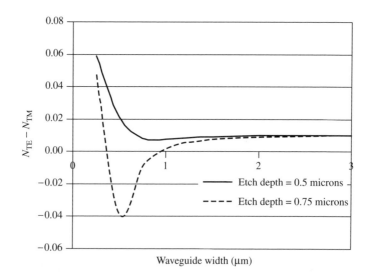

Figure 7.15 Variation of fundamental-mode effective index difference of a 1 μm waveguide (height) for variation in etch depth and waveguide width (beam propagation method)

height. Thus it is probably easier to fabricate a large rib waveguide that is polarisation-independent than a small one. It is also worth noting that the difference in effective indices of the wider of the larger ribs (to the right-hand side of Figure 7.14) saturates at a level of less than 10^{-4}. This is very much smaller than the equivalent value in Figure 7.15, which is approximately 10^{-2}, two orders of magnitude larger. Hence larger ribs are inherently less birefringent than smaller ones. This is because the relative effect of the waveguide boundary is enhanced in smaller waveguides.

7.5 THE EFFECT OF STRESS

It is well known that mechanical stress in a material can induce changes in the optical properties of that material. The existence of stress within a silicon wafer can turn a nominally isotropic material into an optically anisotropic one. This is known as *stress birefringence*, or *the photo-elastic effect*. The semiconductor processing laboratory, together with device packaging, offer numerous opportunities of introducing stress to semiconductor devices and wafers. For optical waveguides this means that the potential exists for stress-induced changes in refractive index of the waveguiding layer. For example, the apparently simple act of depositing a protective oxide layer, or a metallic contact, may introduce stress into a silicon-based optical system. This is particularly problematic when the resulting strain field is directional, because this will result in directional changes in the optical properties of the waveguide. This commonly translates to polarisation-dependent changes in propagation characteristics and/or losses.

Of course, these 'accidental' stress fields will typically be of arbitrary direction, but the near planar surface of a silicon circuit means that it is relatively easy to accidentally introduce a directional stress field.

Rather than attempt to quantify some arbitrary directional change in polarisation characteristics of a silicon waveguide, it is perhaps instructive to consider an application where stress in an optical system is made a virtue. Such a case is the high-birefringence optical fibre (also called the 'polarisation maintaining fibre'). In such fibres, a directional stress field is deliberately introduced to ensure that orthogonal polarisation modes propagate with significantly different propagation constants. That is to say the fibre is birefringent. We have seen in Chapter 2 that a circularly symmetric optical fibre results in orthogonal polarisations propagating with essentially the same propagation constant. This means power transfers between these modes, resulting in an output wave with a random

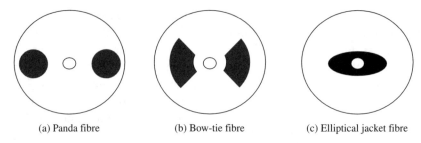

(a) Panda fibre   (b) Bow-tie fibre   (c) Elliptical jacket fibre

Figure 7.16 Three types of high-birefringence fibre formed by stress-inducing inclusions

state of polarisation. If the propagation constants of orthogonal modes are sufficiently different, little or no such exchange of power results, and hence the state of polarisation is maintained.

Figure 7.16 shows three examples of polarisation-maintaining fibres fabricated with stress-generating inclusions. The shaded areas represent doped silica regions that have large thermal expansion coefficients, which are inserted to introduce directional stress when cooled after fibre pulling.

Typically a tensile force is generated in one cross-sectional direction, and a compressive force in the orthogonal direction. This changes the refractive indices by the photoelastic effect. For a polarisation-maintaining fibre, the modal birefringence, B, is usually quantified in terms of the normalised difference of the orthogonal propagation constants:

$$B = \frac{\beta_x - \beta_y}{k_0} \tag{7.12}$$

Typical modal birefringence for a polarisation-maintaining fibre is of the order of $B = 5 \times 10^{-4}$.

7.6   DISCUSSION

It is clear from the foregoing sections that the response to light of differing polarisation can be vital to the operation of a silicon optical circuit. In particular when the circuit is fed by a standard communications optical fibre, resulting in excitation of the circuit with randomly polarised light, in some applications it is important that the circuit be insensitive to polarisation. We have seen the effect of polarisation in propagation in planar and rib waveguides and in coupling to such waveguides, the effect of multimode waveguides, and the effect on some simple devices. Prior to discussing the entire effect of polarisation, it is perhaps worth considering one further specific effect on a device, a device that is particularly affected by relative phase, the arrayed waveguide grating (AWG).

7.6.1   The Effect of Polarisation and Multimode Sections on the AWG

In Chapter 6, the AWG as a wavelength demultiplexer was discussed. A series of discrete coherent sources were used to demonstrate that a linearly increasing phase shift across a number of sources allows multibeam interference to result in a focusing effect, and constructive interference at a point. This is used to advantage in the AWG, in which the total phase change accumulated as a wave passes through the device is used to allow multibeam interference to occur in an output slab waveguide. Consequently, there are two specific effects discussed in this chapter that may compromise the effectiveness of the AWG to successfully allow constructive interference in the output slab; these are (i) the multimode behaviour of the input and output slabs, and (ii) the polarisation sensitivity of the entire device. In both cases the existence of modes with differing propagation constants (either higher-order modes or modes of different polarisation) will result in the AWG focusing at a slightly different position in the output slab waveguide of the AWG. This appears as an apparent frequency shift in the output of the device, and effectively results in channel crosstalk in the AWG. The frequency shift, Δf, between polarisation modes has been conveniently described by Smit and van Dam [7] as:

$$\Delta f \approx f \frac{N_{TE} - N_{TM}}{N_{TE}^{g}} \qquad (7.13)$$

where f is the centre wavelength of the AWG, N_{TE} and N_{TM} are the effective indices for the TE and TM polarisations, and N_{TE}^{g} is the group index of the waveguide TE mode.

Let us consider the effect of the multimode behaviour of the input and output slabs, and the polarisation sensitivity of the AWG in turn.

Multimode Behaviour of the Input and Output Slabs

We have seen in this chapter that different modes of a planar waveguide will propagate with different resulting phase shifts. This will result in 'ghosting' in the output slab of the AWG, as each mode has a different focal point in the output slab. In effect this is additional crosstalk. It should be noted, however, that this is not a phenomenon unique to SOI AWGs, but will occur in all AWGs that employ multimode slab regions within the device. The degree of crosstalk will be related to the phase differential between modes, as calculated earlier, but crucially, also to the relative optical power in each of the waveguide modes. Consequently, even if higher-order modes exist, if power can be prevented from coupling to these modes, or the modes can be made to be lossy, then little ghosting will occur.

In order to estimate the relative coupling from rib waveguides to the modes of the slab regions, an overlap integral must be performed. This has been carried out, for example, by Pearson et al. [8], who showed that for a silicon on insulator AWG based upon a 1.5 μm silicon layer, coupling to the fundamental mode of the slab is more than two orders of magnitude higher than coupling to higher-order modes. This is demonstrated in Figure 7.17.

This demonstrates that it is possible to design an AWG such that the multimode nature of the slab region does not unduly affect the output interference pattern, because higher-order modes of the slab are very poorly excited. Consequently the polarisation dependence of the AWG is much more likely to be problematic than the modal characteristics.

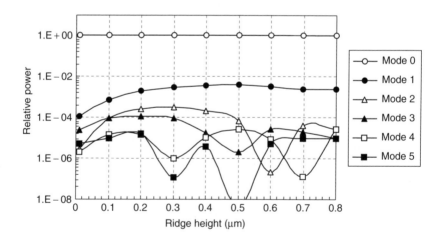

Figure 7.17   Relative power coupling to slab modes from a single-mode input rib. Source: M R T Pearson, A Bezinger, A Delage et al. 'Arrayed waveguide grating demultiplexers in silicon-on-insulator', *Proc. SPIE*, **3953**, 2000. Reproduced by permission of SPIE

Polarisation Sensitivity of the AWG

The polarisation sensitivity of the AWG arises from two sources, the polarisation dependence of the array waveguides, and the polarisation dependence of the slab regions. Both these effects have been discussed in preceding sections of this chapter, and hence will be discussed only briefly here.

Firstly we have seen that the array waveguides (rib waveguides) will, in general, be polarisation-dependent. However, we have also seen that it is possible to design such waveguides to be nominally polarisation-independent. In practice this design may be compromised somewhat by unwanted strain, although even this could be 'designed out' if consistent effects were present.

We have not, however, considered compensating the polarisation dependence of the slab region of the AWG. Cheben et al. [9] studied this problem, producing a partially etched region of both the input and output slabs. The aim was to totally neutralise the phase difference between TE and TM modes in the AWG (in both slab and rib regions), by creating a prism-like region in the slabs, thus effectively altering the optical path lengths of the TE and TM modes in the slab regions, to compensate for the relative phase differences. This is an extremely elegant solution to the problem, producing impressive results. A graph relating etch depth of the compensating region and polarisation dispersion for both theoretical and experimental results is shown in Figure 7.18.

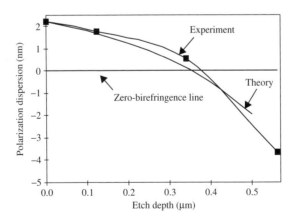

Figure 7.18 Dispersion of the compensator with etch depth. Source: P Cheben, A Bezinger, A Delage et al. 'Polarisation compensation in silicon-on-insulator arrayed waveguide grating devices', *Proc. SPIE*, **4293**, (2001). Reproduced by permission of SPIE

A variety of other approaches to polarisation compensation have also been reported. In 1993, Zirngibl et al. [10], designed an InP AWG with a free spectral range to be the same as the separation of the TE and TM mode patterns in the output slab. This means that the grating order of the AWG for the TE mode is exactly one order higher than for the TM. Thus the two patterns are coincident in the output slab. The disadvantage of this method is that the spectral width of the device is limited by the polarisation dispersion, resulting in a relatively narrowband device.

A more elegant solution was demonstrated by Takahashi et al. [11], who inserted a quartz $\lambda/2$ plate into the mid point of their AWG. The halfwave plate converts TE polarised light to TM, and vice versa. Hence light entering the AWG as TE polarised light propagates through half the device as TE polarisation, and half the device as TM polarisation. Similarly light entering as TM polarisation propagates the second half of the device as TE polarisation. Consequently exactly the same phase change is experienced by the TE and TM modes. Of course the assumption is that the halfwave plate can be placed exactly at the mid point of the AWG. This approach is suitable only if a halfwave plate can be used that is small compared to the divergence of the waveguides, since the 'gap' containing the halfwave plate must be crossed without too much loss. In silicon waveguides the numerical aperture is large owing to the large refractive index contrast, and hence this approach is problematic.

A method similar to that of Cheben et al. [9] (above) was demonstrated by Zirngibl et al. [12]. In this case the compensating region was placed in the waveguide array, rather than in the slabs. A region of waveguide is introduced into each array waveguide, with a different birefringence from that of the main part of the waveguide. Thus the phase difference, $\Delta\phi$, between adjacent waveguides is given by:

$$\Delta\phi = k_0[N_1 \Delta L + \delta L(N_1 - N_2)] \qquad (7.14)$$

where ΔL is the length difference between adjacent waveguides, δL is the additional length of compensating waveguide, N_1 is the effective index of the original waveguide, and N_2 is the effective index of the compensating section of waveguide. If ΔN_1 and ΔN_2 are the differences between the TE and TM values of N_1 and N_2, the condition for $\Delta\phi = 0$ is found to be:

$$\frac{\delta L}{\Delta L} = \frac{1}{\left(\dfrac{\Delta N_1}{\Delta N_2} - 1\right)} \qquad (7.15)$$

Hence the entire waveguide array can be made polarisation-independent by inserting compensating sections in the array waveguides with increasing multiples of length δL.

One other compensation technique reported in the literature is worth mentioning. This is the addition of a polarising splitter at the input to the AWG [7]. This is of course a standard technique for eliminating birefringence, but in most devices it results in duplication of the circuit in question. In an AWG, the situation is simplified, because rather than duplicating the entire AWG, it is sufficient simply to use different input waveguides for the TE and TM modes. If the input separation is chosen to be equal to the output shift between the polarisations, then the TE and TM outputs will be coincident in the output slab. Of course this configuration is suitable only if the AWG use is to be limited to the demultiplexing function. Applications such as $N \times N$ routing, which use multiple inputs, are obviously excluded.

7.6.2   The Effect of PDL on Other Devices

The preceding sections have concentrated the discussion of PDL on the AWG, although we have seen that the diversity of effects described under the global term of PDL will affect different devices in different ways. The effect on the AWG is dominated by considerations of relative phase. This will also be true of any interferometric device. However, any loss-limited device will be affected more by the relative loss of the TE and TM polarisations. For example, a receiver working close to its sensitivity limit will be differentially affected by the loss of the TE and TM modes. Consequently, issues such as surface scattering, and the associated polarisation-dependent loss, are important in this situation. Similarly, circuits with waveguides with a significant amount of bending may be differentially affected by the different bending loss of the waveguides. Alternatively, evanescent-based devices such as couplers will be affected by the different degrees of confinement of the two polarisations. Hence it is clear that the device-specific effects of polarisaton must be considered when attempting to evaluate the overall impact of polarisation on any given device.

7.7   CONCLUSION

This chapter has discussed the effect on devices of different responses to light polarised in the TE or TM directions. In particular the effect of waveguide dimensions, surface scattering, polarisation-dependent coupling loss, waveguide birefringence, and phase compensation techniques have been discussed and evaluated. Occasionally a specific device has been used to demonstrate or quantify the effects of polarisation.

It is clear from the details contained within this chapter that polarisation of the incoming signal can have a dramatic impact upon the performance of a given optical circuit or device, particularly when that device is sensitive to variations in phase. The overriding conclusion must be that, in order to evaluate the effects of polarisation, the dominant contributions to the polarisation dependence must be identified and quantified. It is also clear from the foregoing that much can be done to combat the effects of polarisation in silicon-on-insulator photonics. Indeed, we have demonstrated that, in principle, many of the effects of polarisation can be eliminated entirely.

However, it is also clear that minimising the effects of polarisation of one parameter may increase the effects of another. For example, if birefringence is removed by waveguide design, the TE and TM mode shapes will still be different. This may increase differential loss via interface scattering. Similarly if birefringence is removed in a small waveguide by selecting a specific waveguide geometry, this may result in a multimode waveguide, which could in turn degrade the performance of an interferometer, or result in loss due to leakage from higher-order modes. Thus the parameter to be optimised must not be considered in isolation, but as a part of an integrated design process.

REFERENCES

1. P K Tien (1971) 'Light waves in thin films and integrated optics', *Appl. Optics*, **10**, 2395–2413.
2. D Marcuse (1972) *Light Transmission Optics*, Van Nostrand Reinhold, New York, ch. 9.
3. J P R Lacey and F P Payne (1990) 'Radiation loss from planar waveguides with random wall imperfections', *IEE Proc. Optoelectron.*, **137**, 282–288.
4. F Ladouceur, J D Love and T J Senden (1994) 'Effect of sidewall roughness in buried channel waveguides', *IEE Proc. Optoelectron.*, **141**, 242–248.
5. M Born and E Wolf (1999) *Principles of Optics*, 7th edn, Cambridge, Cambridge University Press, ch. 15.
6. A Rauber (1978) 'The chemistry and physics of lithium niobate', in E Kaldis (ed.), *Current Topics in Materials Science* 1, North-Holland, Amsterdam, ch. 7.
7. M K Smit and C Van Dam (1996) 'PHASAR-based WDM-devices: principles, design, and applications', *IEEE J. Quantum Electron.*, **2**, 236–250.

8. M R T Pearson, A Bezinger, A Delage et al. (2000) 'Arrayed waveguide grating demultiplexers in silicon-on-insulator', *Proc. SPIE*, **3953**, 11–18.

9. P Cheben, A Bezinger, A Delage et al. (2001) 'Polarisation compensation in silicon-on-insulator arrayed waveguide grating devices', *Proc. SPIE*, **4293**, 15–22.

10. M Zirngibl, C H Joyner, L W Stulz, Th Gaiffe and C Dragone (1993) 'Polarisation independent 8×8 waveguide grating multiplexer on InP', *Electron. Lett.*, **29**, 201–202.

11. H Takahashi, Y Hibino and I Nishi (1992) 'Polarisation-insensitive arrayed-waveguide grating wavelength multiplexer on silicon', *Optics Lett.*, **17**, 499–501.

12. M Zirngibl, C H Joyner and P C Chou (1995) 'Polarisation-compensated waveguide grating router on InP', *Electron. Lett.*, **31**, 1662–1664.

8
Prospects for Silicon Light-emitting Devices

In this final chapter, we move away from descriptions of the physics and technology of demonstrated device operation and turn to a subject which at the time of writing is one of the most hotly researched areas in semiconductor science: the goal to fabricate silicon-based, electrically pumped, light-emitting devices (LEDs) with efficiencies comparable to those obtained from III–V semiconductor materials. Such devices would find application in optical communication on all scales, from intra- and inter-chip interconnects to the fibre-optic network. Further, the technology required for efficient emitters would permit the fabrication of planar optical amplifiers. The hurdles to be negotiated to generate efficient electroluminescence are significant. Silicon is an indirect band-gap semiconductor with a low probability for phonon-assisted, radiative electron–hole recombination (i.e. resulting in the spontaneous emission of photons). Relatively fast, nonradiative recombination mechanisms – such as those via lattice defects or the Auger mechanism – dominate, hence the internal quantum efficiency for silicon luminescence is only 10^{-6}.

The global research effort to create silicon LEDs and ultimately silicon-based lasers is intense (see [1] for a complete review of the state of the art at the time of writing). Here we describe several of the most promising approaches currently under investigation in this exciting area of silicon photonic development.

8.1 ERBIUM DOPING

Rare-earth ions, especially erbium, have played a significant role in the development of the optical communications network. Trivalent erbium

Silicon Photonics: An Introduction   Graham T. Reed and Andrew P. Knights
© 2004 John Wiley & Sons, Ltd   ISBN: 0-470-87034-6

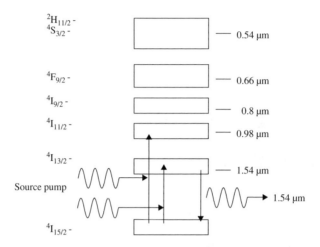

Figure 8.1 Representation of Er^{+3} electronic energy levels. Pumping sources of 980 nm and 1480 nm are shown together with the emission of photons at a wavelength of 1540 nm

ions (Er^{+3}) embedded in semiconductor or dielectric materials have an incomplete 4f electronic shell, permitting intra-4f transitions when excited (or pumped) by an optical source. Further, the first excited state is at an energy of 0.8 eV, hence down-transitions following optical pumping with higher-energy photons (usually from either a 980 nm or 1480 nm laser source) can occur with the emission of photons at a wavelength of 1.54 μm. The energy levels of Er^{+3} are shown schematically in Figure 8.1.

In the 1980s it was shown that erbium-doped optical fibres could be used as optical amplifiers when pumped by a laser source [2]. At that time, the need for efficient amplifiers for the telecommunications network was paramount, and erbium-doped fibre amplifiers (EDFAs) addressed this requirement. Subsequent developments led to widespread deployment of EDFAs, fixing the communication wavelength at 1.54 μm.

The same principle of photon emission as used in fibre amplifiers can be applied to erbium-doped silicon. As we will see in the following section, electrically pumped devices have been formed, using erbium as the optically active component in a silicon matrix, which emit photons at a wavelength of 1.54 μm.

8.1.1 Erbium Ion Implantation

A significant restriction on the fabrication of silicon optical sources is the need for the fabrication methods to be completely compatible with processing techniques used in the manufacture of more standard devices in an ultra-large-scale-integration (ULSI) process flow. Many of the current approaches utilise ion implantation for the formation of the optically active medium. Ion implanters are ubiquitous in silicon device fabrication facilities and devices using implant technology usually meet the requirement for ULSI compatibility. Implantation is a convenient method for introducing erbium into silicon [3], although erbium's high atomic mass restricts the projected range to about 15 % of that for a boron ion of equivalent energy. Therefore, many of the experimental reports in the research literature describe setups using implanters capable of producing ions with energies of >1 MeV.

Erbium doping of silicon progresses in a similar manner to that for more common dopants such as boron. Following ion implantation, a thermal treatment repairs the silicon lattice damage while activating the charged erbium ion. Subsequent fabrication of a $p-n$ junction allows the generation of charge carriers which can recombine via energy transfer to the Er^{3+} ion. Once excited, the Er^{3+} ion can in turn decay with the emission of a 1.54 μm photon (see Figure 8.2). The efficiency limits of this electroluminescence mechanism were described by Xie et al. [4] who concluded that, for performance comparable with InGaAsP/InP devices, researchers should concentrate their efforts on increasing the concentration of active erbium incorporated into silicon.

Ion implantation is a nonequilibrium process and as such can be used to introduce large concentrations of any dopant into the near-surface

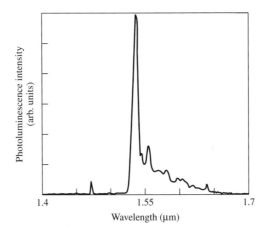

Figure 8.2 Typical luminescence spectrum for erbium-doped silicon. Reproduced from Y H Xie, E A Fitzgerald and Y J Mii (1991) 'Evaluation of erbium-doped silicon for optoelectronic application', *J. Appl. Phys.*, **70**, 3223–3228 by permission of American Institute of Physics

region of a silicon wafer. However, the thermal step required to reconstruct the silicon lattice does not guarantee all of the dopant is incorporated into the silicon matrix in a useful manner. For common electrical dopants the limit of solid solubility is $>5 \times 10^{20}$ cm^{-3}, but for erbium it is only 1×10^{18} cm^{-3}. Polman et al. [5] investigated different mechanisms for optically activating erbium implanted into silicon. Using a low-temperature (600 °C) anneal to reconstruct an amorphous layer created by the Er implantation (a process known as *solid-phase epitaxy*), they were able to incorporate a concentration 1×10^{20} cm^{-3} Er ions. However, they observed a complex interaction of the implanted ions and implant-induced defects, resulting in the segregation of a large proportion of the erbium out of the crystalline structure, thus rendering it of no use in the formation of optically active centres.

A significant discovery related to Er incorporation was the role played by oxygen. In a subsequent report to [5], the same group co-implanted 1-MeV Er$^+$ ions to a dose of 1.6×10^{15} cm^{-2} and 160-keV O ions to a dose of 5×10^{15} cm^{-2}, such that the implantation profiles of Er and O overlapped. Following solid-phase epitaxy of the implanted amorphous layer, almost no segregation was observed with the annealed Er profile being coincident with the as-implanted profile. It appeared that the erbium formed a bond with the oxygen and this so-called complex was less easily segregated from the silicon crystal [3].

8.1.2 Optical Efficiency of Er-implanted Si

After description of erbium incorporation into the silicon crystal our attention is necessarily drawn to luminescent efficiency. Following the recipe for high concentration incorporation described in [5], photoluminescence measurements (i.e. measurement of optical emission following optical excitation) were performed on as-prepared samples and those subjected to a further anneal at 1000 °C for 15 seconds [6]. The results are shown in Figure 8.3.

The spectrum includes peaks resulting from Si band-edge recombination (1.13 μm), and from the de-excitation of Er^{3+} at 1.54 μm. The data show that both the band-edge and Er luminescence can be increased by three and five times respectively by the 1000 °C anneal. This is attributed to the reduction of crystalline defects which are still present following the solid-phase epitaxial growth at 600 °C. Reduction in defect concentration and hence competitive nonradiative recombination, increases the efficiency of the Er^{3+} excitation.

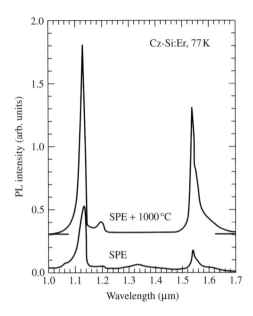

Figure 8.3 Photoluminescence spectra taken at 77 K for Er-implanted Si after solid-phase epitaxial recrystallisation at 600 °C (lower spectrum) and after subsequent thermal annealing at 1000 °C for 15 seconds. Reproduced from A Polman, G N van den Hoven, J S Cluster, J H Shin and R Serna (1995) 'Erbium in crystal silicon: optical activation, excitation, and concentration limits', *J. Appl. Phys.*, **77**, 1256–1262 by permission of American Institute of Physics

The photoluminescence intensity as a function of implanted erbium dose was obtained in the same study and is reproduced in Figure 8.4. It is plotted together with the Er photoluminescence lifetime. These data show that there is a clearly defined limit as to the concentration of optically active erbium that can be incorporated into a silicon device. In the case of [6] this limit was deduced from data fitting to be 3×10^{17} cm^{-3}, much less than the concentration of 1×10^{20} cm^{-3} which is incorporated into the silicon crystal.

The nature of the optically active erbium was shown to be related to the presence of oxygen. The data shown in Figure 8.4 were obtained from silicon samples containing a background concentration of oxygen of 1.7×10^{18} cm^{-3}. Assuming that each optically active Er ion forms a bond with 4–6 oxygen atoms (as suggested by experimental observation), a maximum optically active concentration of between 2–6×10^{17} cm^{-3} can be derived, in close agreement with the measured data. The important conclusion is then that the active limit is determined by the concentration of oxygen co-doping, and not the total amount of embedded erbium.

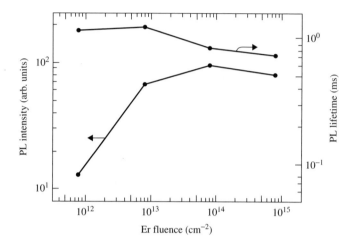

Figure 8.4 Photoluminescence and lifetime at 1.54 μm as a function of Er fluence, measured at 77 K. The pump power at 514.5 nm was 160 mW. Reproduced from A Polman, G N van den Hoven, J S Cluster, J H Shin and R Serna (1995) 'Erbium in crystal silicon: optical activation, excitation, and concentration limits', *J. Appl. Phys.*, 77, 1256–1262 by permission of American Institute of Physics

Further, one can estimate the internal quantum efficiencies for radiative recombination to be 10^{-3} and 10^{-6} for the erbium–oxygen complex and erbium-only centres respectively [3].

8.1.3 Optical Intensity Quenching

A significant problem associated with erbium-doped silicon LEDs is the severe signal quenching observed at room temperature. Figure 8.5 shows a plot of photoluminescence (PL) intensity versus measurement temperature for silicon either implanted with erbium and oxygen, or with just erbium. Post-implantation thermal treatments were similar to those outlined previously [7].

For the erbium-only doped sample a significant reduction in PL intensity is observed as the measurement temperature is increased from 77 K to 300 K. At room temperature, the intensity is so weak as to be unmeasurable. However, the presence of oxygen is found to be beneficial in the reduction of quenching effects and a small amount of luminescence is measured for the co-doped sample. It has been suggested that the most significant contributions to temperature quenching in (Er + O)-doped silicon are via: (1) a 'back-transfer' mechanism in which Er^{3+} ions, excited by trapped carrier recombination, relax via a nonradiative transfer of energy to valence electrons which are in turn excited to a

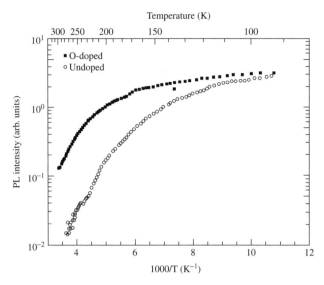

Figure 8.5 Temperature dependence of the photoluminescence intensity at 1.54 μm for both erbium-only doped (referred to in the figure as undoped) and erbium and oxygen co-doped (referred to in the figure as O-doped) samples. Source: S Coffa, G Franzò, F Priolo, A Polman and R Serna 'Temperature dependence and quenching processes of the intra-4f luminescence of Er in crystalline Si', *Phys. Rev. B.*, **49**, 16313–16320, © 1994 by the American Physical Society

defect level located inside the forbidden energy band-gap; and (2) the Auger excitation of free carriers [3].

8.1.4 Electroluminescent (EL) Devices

The creation of a light-emitting diode (LED) requires the fabrication of a *p–n* junction within an optically active medium [8]. In principle, such electroluminescent devices can be made in a straightforward manner using Er- or (Er + O)-doped silicon. One such design was described by Libertino et al. [9] and is shown in Figure 8.6.

When such Si LEDs are conventionally, electrically pumped using forward bias, extremely low efficiencies are observed at room temperature (Figure 8.7) [10]. In particular, there is a strong thermal quenching of the optical power as the temperature is raised towards 300 K.

In EL devices under forward bias, the excitation of the Er ions takes place via the recombination of electron–hole pairs. This recombination suffers from a high probability of competing nonradiative processes in the form of Auger excitation of the high number of free carriers present in the vicinity of the forward-biased junction, and the likelihood of

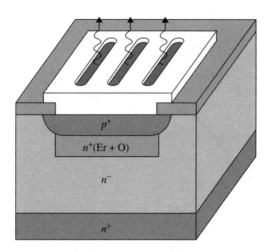

Figure 8.6 Surface emitting (Er + O)-doped Si LED. Reprinted from *Mat. Sci. Semicond. Process.*, **3**, S Libertino, S Coffa and M Saggio, 'Design and Fabrication of integrated Si-based optoelectronic devices', 375–381, © 2000, with permission from Elsevier

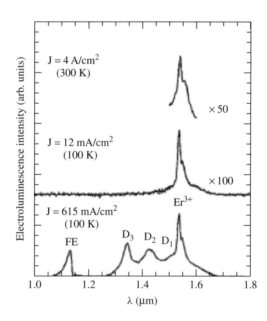

Figure 8.7 Electroluminescence spectra of a surface-emitting Si:Er LED for different drive current densities at 100 K and at room temperature. Reproduced from B Zheng, J Michel, F Y G Ren et al. (1994) 'Room-temperature sharp line electroluminescence at $\lambda = 1.54\,\mu$m from an erbium-doped, silicon light emitting diode', *Appl. Phys. Lett.*, **64**, 2842–2844 by permission of American Institute of Physics

recombination at residual lattice defects remaining from the implantation of the Er ions during the fabrication of the device.

Following earlier work, Franzò et al. [11] discussed methods for the fabrication and optimisation of electrically pumped (Er + O)-doped Si LEDs. They summarised the important role of oxygen in four areas (as discussed in sections 8.1.1–8.1.3):

1   Oxygen increases the effective solubility of Er in Si

2   It inhibits Er segregation during solid-phase epitaxy

3   It enhances luminescent yield

4   It reduces the temperature quenching of the luminescence.

They also showed that reverse-biasing the devices to the point of junction breakdown (in their case >5 V applied) produces a far more efficient method (compared to forward-biasing) for generating electroluminescence in (Er + O)-doped Si LEDs (Figure 8.8). Notice that the reverse-biased signal is reduced in the figure by a factor of 16 for convenient

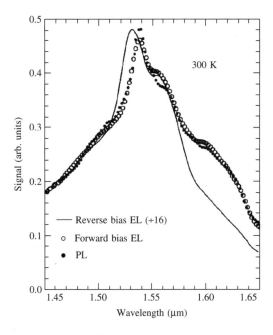

Figure 8.8 Room-temperature EL signal under both forward and reverse bias operation at a current density of 2.5 A·cm^{-2}. Reproduced from G Franzò, S Coffa, F Priolo and C Spinella (1997) 'Mechanism and performance of forward and reverse bias electroluminescence at 1.54 μm from Er-doped Si diodes', *J. Appl. Phys.*, **81**, 2784–2793 by permission of American Institute of Physics

comparison with the forward-biased signal. In the case of reverse-bias junction breakdown, Er ions are excited by impact ionisation. Because the optically active (doped) region is inside the depletion layer of the junction there are no free carriers, so the Auger decay mechanism is severely inhibited. In fact, Franzò measured an Er decay lifetime of over $100 \, \mu s$, indicating that all fast, nonradiative mechanisms are suppressed in the reverse-biased diodes.

The fabrication of reversed-biased Si LEDs was a significant step forward towards the goal of efficient electroluminescence. We shall return to the same excitation mechanism in section 8.2.4.

8.2 LOW-DIMENSIONAL STRUCTURES

The indirect band-gap of silicon and corresponding low probability for radiative recombination is the main reason why nonradiative transitions dominate the relaxation of excited carriers. One way to prevent carrier diffusion to nonradiative centres during their relatively long lifetime is to confine them in low-dimensional structures.

8.2.1 Porous Silicon

In 1990, Canham [12] published a pioneering paper demonstrating efficient room-temperature luminescence from silicon samples which had been exposed to an anodisation process. Specifically, silicon was placed in an electrochemical cell consisting of hydrofluoric acid through which a current of several milliamps was passed. The resulting structure of the samples consisted of an array of small holes that ran orthogonally to the sample surface. The dimension of the holes was shown to be controlled by the anodising conditions. If the holes were made so large as to overlap, isolated pillars formed on the sample surface. Following exposure to aqueous HF acid, this so-called porous silicon exhibited efficient room-temperature photoluminescence (PL) upon irradiation with Ar laser lines at 488 and 514.5 nm (Figure 8.9).

The efficient PL is a direct result of the small width of the silicon pillars. As the pillars are reduced in dimension to below a few nanometres, quantum confinement of the excited carriers results in an effective enlargement of the silicon band-gap, increasing the probability of recombination. In addition, the localisation of the carriers prevents their diffusion to possible nonradiative recombination centres, thus increasing the chances of radiative down-transitions.

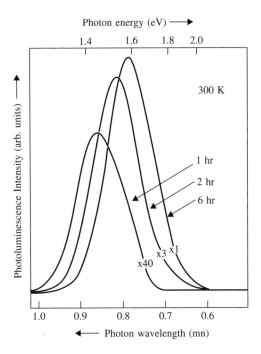

Figure 8.9 Room-temperature photoluminescence from anodised p-type silicon following subsequent immersion in aqueous HF acid. The optical pump consisted of 200 mW at a wavelength of 514.5 nm. Reproduced from L T Canham (1990) 'Silicon quantum wire array fabrication by electrochemical and chemical dissolution of wafers', *Appl. Phys. Lett.*, **57**, 1046–1048 by permission of American Institute of Physics

The first electroluminescent porous silicon device was demonstrated by Richter et al. [13]. Following the fabrication of a 75-µm thick porous layer in an n-type silicon wafer, a thin film (12 nm) of gold was deposited on the porous silicon layer. After the formation of an ohmic contact to the reverse of the wafer, a current of 5 mA was passed through the device. A voltage of 200 V was required to achieve this modest current flow. A small but observable emission (with the eye in a darkened environment) was recorded and is shown in Figure 8.10. Efficient photoluminescence from a similarly prepared sample is shown for comparison.

Considerable progress in the fabrication of EL porous silicon LEDs is evidenced by a recent paper of Gelloz and Koshida [14]. In it they describe the fabrication of a 1-µm thick porous silicon layer, subsequently oxidised using an electrochemical process. Indium tin oxide is used as the electrode material for the porous silicon layer with aluminium providing an ohmic contact to the wafer backside. Room-temperature

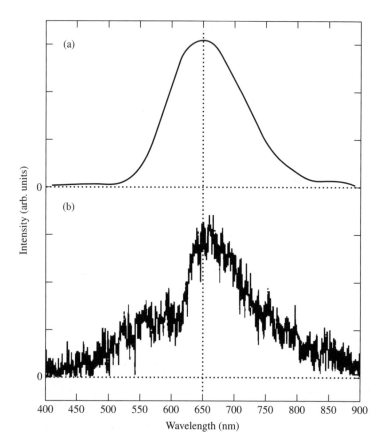

Figure 8.10 Spectra of light-emitting porous silicon pumped (a) optically and (b) electrically. Source: A Richter, P Steiner, F Kozlowski and W Lang (1991) 'Current-induced light emission from a porous silicon device', *IEEE Electron. Device Lett.*, **12**, © 2003 IEEE

EL with an external quantum efficiency approaching 1 % for an applied bias of 5 V is reported, although the exact nature of the carrier excitation process is not known. These devices are the most efficient porous Si LEDs reported to date (Figure 8.11).

Since Canham's original paper there has been an enormous interest in porous silicon research (for a comprehensive review see [15]). One reason for the plethora of published work is the low-cost fabrication method available even for those with the most modest research budgets. Indeed this cost-effective fabrication route is an advantage of porous silicon, although concerns exist as to the compatibility of porous silicon with

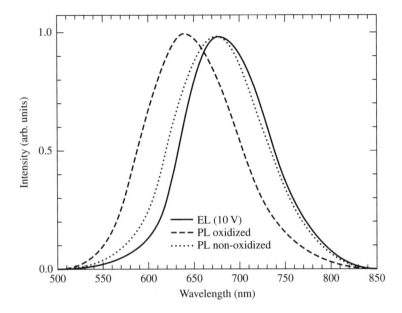

Figure 8.11 Electroluminescence at an applied voltage of 10 V for porous silicon, anodically oxidised LEDs. Normalised photoluminescence spectra for similar structures is shown for comparison. Reproduced from B Gelloz and N Koshida (2000) 'Electroluminescence with high and stable quantum efficiency and low threshold voltage from anodically oxidised thin porous silicon diode', *J. Appl. Phys.*, **88**, 4319–4324 by permission of American Institute of Physics

more mainstream silicon processing. A further issue with this technology is the emission wavelength which is restricted to a range associated with sub-band-gap energies. This makes porous silicon unsuitable for use as a signal source for long-haul telecoms, but does not preclude it as a potential pump source or for communication over short distances.

8.2.2   Nano-crystals

An approach that combines the advantages of low-dimensional silicon with a robust ULSI-compatible fabrication process is the use of silicon nano-crystals (Si-nc) embedded in a dielectric matrix, most commonly SiO_2. A straightforward fabrication technique involves the formation of a silicon-rich, sub-stoichiometric $SiO_x (x < 2)$ film on a silicon substrate, followed by a high-temperature anneal in the region of $1200\,°C$ for several minutes. The thermal energy promotes the phase separation of

Si and SiO_2 and the final structure consists of small silicon nano-crystals whose size and distribution depend on the original film properties and the subsequent thermal anneal.

Iacona et al. [16] produced silicon-rich films by plasma-enhanced chemical vapour deposition (PECVD) with x varying between 1 and 1.75. The thermal treatment consisted of an anneal at between 1000 °C and 1300 °C for one hour. The resulting Si-nc structures were reported to have a mean radius of between 0.7 and 2.1 nm. Photoluminescence measurements following excitation with a 488-nm Ar laser are reproduced in Figure 8.12. The authors exhibited repeatable control

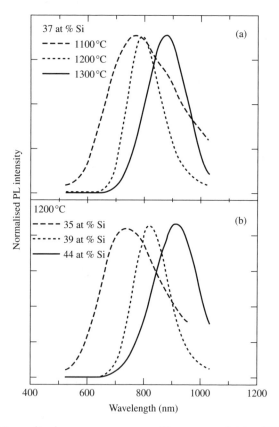

Figure 8.12 Normalised room-temperature PL spectra of (a) a SiO_x film having a Si concentration of 37 % after thermal annealing processes performed at 1100, 1200 and 1300 °C for 1 hour; and (b) SiO_x films having Si concentrations of 35, 39 and 44 % after thermal annealing performed at 1200 °C for 1 hour. Reproduced from F Iacona, G Franzò and C Spinella (2000) 'Correlation between luminescence and structural properties of Si nanocrystals', *J. Appl. Phys.*, **87**, 1295–1303 by permission of American Institute of Physics

over both the emission wavelength and the PL intensity, with the peak wavelength shifting from 650 nm to 900 nm as the percentage of silicon in the deposited film was increased from 35 % to 44 %. Annealing at 1250 °C produced the greatest PL intensity, with an increase of nearly two orders of magnitude compared to samples annealed at 1000 °C.

A different method for embedded Si nano-crystal fabrication was demonstrated by Min et al. [17]. They implanted Si^+ ions into a 100-nm thick thermally grown SiO_2 film. The implantation energy was 50 keV and the doses ranged from 1 to $5 \times 10^{16} \, cm^{-2}$. The implanted samples were annealed at temperatures in the range 1000–1200 °C for 10 minutes. In addition to demonstrating efficient PL, the authors were able to address issues relating to the origin of the luminescence signal. By comparing results from Si^+ and Xe^+ implanted SiO_2 films they distinguished signals emanating from the Si-nc and those originating from SiO_2 defects. This study also highlighted the advantages of using ion implantation to form the nano-crystals; namely the precise control of excess Si concentration and the concentration profile of the formed Si-nc structures, allowing fabrication of samples with predictable and repeatable luminescence characteristics.

Devices in which the optically active medium is insulating (as is the case for SiO_2) would seem incompatible with electrical excitation. Indeed the fabrication of LEDs using this medium has proven troublesome. Even so, by utilisation of a current tunnelling between the embedded nano-crystals, efficient electroluminescence has been demonstrated. In these devices it is likely that control of the Si-nc structure dimensions is compounded by the need for sufficient current densities (dependent on the Si-nc distribution) to create efficient EL. Irrera et al. [18] fabricated a Si-nc LED using silicon PECVD-deposited SiO_x, subsequently annealed at 1100 °C for 1 hour. Contact to the Si-nc containing layer was made via a poly silicon/aluminum stack. For a graphical representation of a device based on this design see Figure 8.13.

The EL obtained from this structure is reproduced in Figure 8.14. It is assumed that the luminescence results from electron–hole recombination inside the Si-nc, initially excited by impact ionisation from hot electron injection (somewhat similar to the mechanism used for efficient EL in (Er + O)-doped reverse-biased Si LEDs described in section 8.1.4).

Finally in this section, it is appropriate to highlight the ground-breaking work of the group lead by Lorenzo Pavesi from which optical amplification was demonstrated raising expectations that Si nano-crystals may provide a route to a silicon laser. The original report of optical gain was published in the journal *Nature* in 2000 [19] but

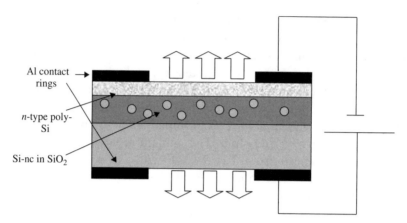

Figure 8.13   Graphical representation of a possible design for a Si-nc LED

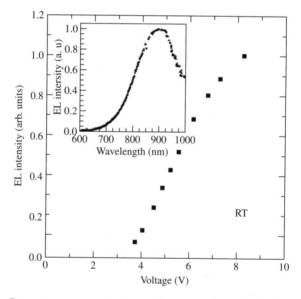

Figure 8.14   Room-temperature EL as a function of applied voltage for a Si-nc LED. Inset is the EL spectrum. Reprinted from *Physica E*, **16**, A Irrera, D Pacifici, M Miritello et al. 'Electroluminescence properties of light emitting devices based on silicon nanocrystals', 395–399, © 2003, with permission from Elsevier

more recently Pavesi presented a complete review of all current attempts to fabricate a silicon laser device [20]. The microscopic details of the gain mechanism is currently under debate; however, Pavesi suggests that it is related to the recombination at localised states, probably related to silicon–oxygen bonds formed at the interface of the Si-nc and the oxide, or within the oxide matrix itself.

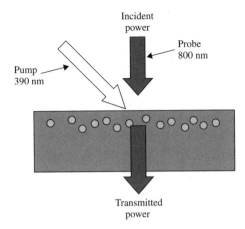

Figure 8.15 Pump–probe experimental setup to measure optical gain of Si-nc

The confirmation of gain in Si-nc has been shown by a number of researchers. In their original experimental report, Pavesi et al. used a straightforward pump–probe arrangement shown schematically in Figure 8.15. The sample was prepared using the method of high-dose $(1 \times 10^{17}\,\text{cm}^{-2})\text{Si}^+$ ion implantation of a high-purity quartz, followed by an anneal of 1100 °C for 1 hour. The use of quartz (as opposed to a silicon substrate) was required because of the transmission geometry of the experiment. The sample was pumped by an intense laser beam at 390 nm, strong enough to provide the population inversion conditions required for optical gain (mean power of $2\,\text{kW/cm}^2$). A relatively weak probe signal of 800 nm was passed through the layer containing the Si-nc with the difference between the incident and output power being recorded. The results are reproduced in Figure 8.16. Significant gain is observed in the region of $10,000\,\text{cm}^{-1}$. Further indication of gain was inferred from measurements of optical loss for power densities $<1\,\text{kW/cm}^2$ (i.e. where population inversion is not achieved), and by the absence of gain for probe wavelengths far away from 800 nm (where the probe is no longer resonant with the optical transition).

8.2.3 Nano-crystals with Erbium

Silicon nano-crystal (Si-nc) technology shows great promise for the development of Si light-emitting diodes and optically pumped amplifiers, with the possible application to a silicon laser structure. However, the operating wavelengths of these devices is in a window spanning 750–1000 nm. For use in the telecoms network it is desirable for light sources to operate at 1.3 μm or 1.5 μm. As we have seen in section 8.1,

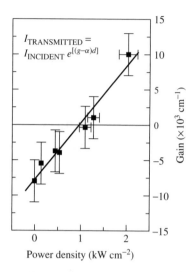

Figure 8.16 Dependence of Si-nc material gain versus pump power density for the experimental setup shown in Figure 8.15. Reprinted by permission from Nature, L Pavesi, L Dal Negro, G Mazzoleni, G Franzò and F Priolo (2000) 'Optical gain in silicon nano-crystals', *Nature*, **408**, 440–444, © (2000) Macmillan Publishers Ltd. www.nature.com/nature

significant advances have been reported in the fabrication of 1.5-μm emitting devices following the incorporation of erbium into silicon. It is therefore a natural extension to combine Si-nc technology with Er doping.

Franzò et al. [21] implanted Er^+ ions into a Si-nc matrix formed by PECVD deposition of silicon-rich films and subsequent annealing at 1250 °C. Following implantation the films were further annealed at a temperature of 900 °C to reduce the implant-induced defect concentration. They found that not only did the Er ions produce luminescence, but the process was more efficient by a factor of 100 compared to (Er + O)-implanted bulk silicon. Their results are reproduced in Figure 8.17. The authors suggested that the Er ions are pumped by electron–hole pairs generated within the Si-nc, but the ions themselves are located in the SiO_2 matrix. The Er luminescence thus benefits from the advantages of Si-nc compared to bulk silicon (enhanced excitation) and the use of a dielectric medium (strongly reduced nonradiative recombination and negligible temperature quenching).

Recently, the group lead by Salvatore Coffa at STMicroelectronics [22] published a report of electroluminescence from Er-doped SiO_2 as a function of concentration of Si-nc. The device processing is compatible with a standard silicon device fabrication route consisting of oxidation,

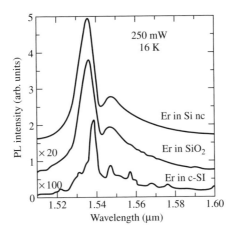

Figure 8.17 Photoluminescence spectra measured at 16 K by pumping with a laser power of 250 mW for three different samples: (Er + O)-implanted crystalline Si; Er-implanted SiO_2; and Er-implanted Si nano-crystals. The erbium concentration is 10^{20} cm^{-3} in all three cases. Reproduced from G Franzò, D Pacifica, V Vinciguerra, F Priolo and F Iacona (2000) 'Er3+ ions-Si nanocrystals interactions and their effects on the luminescence properties', *Appl. Phys. Lett.*, **76**, 2167–2169 by permission of American Institute of Physics

ion implantation, poly-silicon deposition, annealing and metallisation, and results in a device similar in structure to that shown in Figure 8.13.

A tunneling current of 100 μA was used to induce room-temperature EL with an external quantum efficiency of 10 % in a device where the Er-containing SiO_2 was thermally grown (Figure 8.18). The maximum output was found to be limited only by the density of Er ions incorporated into the oxide film (approximately 1×10^{19} cm^{-3}).

Although these Er-doped MOS devices exhibited efficiency approaching that of a commercial III–V LED, they were shown to be limited by their reliability and operating lifetime. This resulted from the use of a tunnelling current through the thermally grown SiO_2 to excite the Er^{3+} ions. By replacing the thermal oxide with a silicon-rich PECVD-deposited SiO_x layer ($x < 2$) and subsequently thermally treating it to form Si-nc before Er ion implantation, Coffa's team was able to significantly improve device reliability. This was explained as resulting from a Si-nc mediated current passing through the SiO_x layer as opposed to a more destructive oxide tunnelling current. However, correlated with increasing reliability was a decrease in the EL efficiency. Even so, acceptable long-life performance was achieved using samples with a small concentration of Si-nc (refractive index of 1.6 as opposed to 1.46 for pure SiO_2), showing an external quantum efficiency of 1 %.

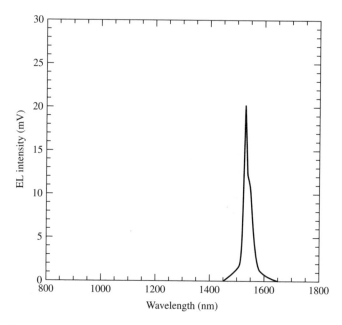

Figure 8.18   Room-temperature EL spectrum measured at 100 μA on a metal-oxide–semiconductor (MOS) device with Er-doped stoichiometric oxide. Reprinted from *Physica E*, **16**, M E Castagna, S Coffa, M Monaco et al. 'Si-based materials and devices for light emission in silicon', 547–553, © 2003, with permission from Elsevier

8.3   DISLOCATION-ENGINEERED EMITTERS

A novel approach to the fabrication of Si LEDs, which is completely compatible with a conventional ULSI process flow, was recently described by researchers at the University of Surrey led by Homewood and Gwilliam [23]. Their reported fabrication process is designed to prevent excited carriers from diffusing to nonradiative recombination centres before radiative transitions can take place. However, unlike the carrier confinement achieved in low-dimensional silicon structures, the Surrey group's design relies on the creation of crystalline defects to provide local three-dimensional carrier confinement. Using an *n*-type silicon wafer substrate, boron is implanted at an energy of 30 keV to a dose of 1×10^{15} cm^{-2}. A conventional anneal step of 1000 °C for 20 minutes is used to electrically activate the implanted boron and repair the primary implantation-induced defects. During the anneal step, displaced silicon atoms, substituted for boron atoms, form dislocation loops at the end of the implanted ion range. The concentration and distribution of the loops is dependent on both the implanted ion dose and the subsequent thermal

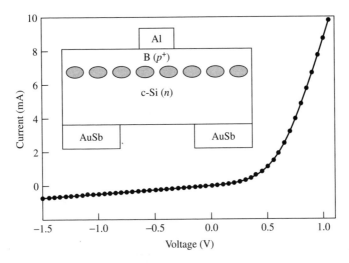

Figure 8.19 Current/voltage plot for a dislocation-engineered Si LED measured at room temperature. The light is emitted through the bottom contact window. Reprinted by permission from Nature, W L Ng, M A Lorenço, R M Gwilliam et al. 'An efficient room-temperature silicon-based light emitting diode', *Nature*, **410**, 192–194, © (2001) Macmillan Publishers Ltd. www.nature.com/nature

processing. Device fabrication is completed with front- and back-side wafer metallisation, with patterned windows allowing the escape of light. A schematic of the device is shown in Figure 8.19 together with the current/voltage characteristic. These data provide clear evidence that these devices exhibit excellent electrical performance, as one would expect from a process that is used routinely in the fabrication of $p-n$ junctions in mainstream silicon device manufacture. This is a significant advantage over those technologies dependent on a highly resistive optically active medium because it ensures efficient carrier injection.

Figure 8.20 shows the EL spectra of this silicon LED. Unlike the (Er + O)-doped Si LEDs, described in section 8.1.4 these devices are driven by a forward bias current. The most striking feature is the apparent increase in total integrated power as the temperature of device operation is raised towards 300 K. This complete lack of temperature quenching was attributed solely to the confinement of the excited carriers and the subsequent prevention of nonradiative recombination via lattice defects.

In a later publication, the same group explored the co-doping of these dislocation-engineered LEDs with Er and beta-phase iron disilicide (β-FeSi$_2$) [24]. The expressed aim was to fabricate LEDs with emission at a wavelength in the region of 1.5 μm. Although the authors conceded that the co-doping process was far from optimised, they demonstrated

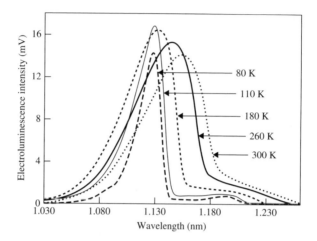

Figure 8.20 Spectra of the electroluminescence intensity versus wavelength at various temperatures. The device was operated at a forward current of 50 mA for all temperatures. Reprinted by permission from Nature, W L Ng, M A Lorenço, R M Gwilliam et al. (2001) 'An efficient room-temperature silicon-based light emitting diode', *Nature*, **410**, 192–194, © (2001) Macmillan Publishers Ltd. www.nature.com/nature

room-temperature EL at a wavelength of approximately 1.55 μm for both Er and β-FeSi$_2$ samples. Control samples which were free from dislocation loops failed to yield any such EL, thus confirming the partial elimination of temperature quenching.

The potential application and widespread deployment of the dislocation-engineered Si LEDs provides an interesting footnote to the search for an efficient silicon light source. The fabrication processes use implantation and anneal recipes which are common in the silicon microelectronics industry. It is likely, then, that tens of millions of dislocation-engineered Si LEDs have in fact been fabricated in the last few decades and are operating in integrated circuits worldwide.

8.4 RAMAN EXCITATION

We have thus far concentrated on the search for material systems exhibiting efficient electroluminescence for the fabrication of silicon-based LED and laser sources. There also exists a requirement for silicon waveguide optical amplifiers which can perform the same function as an EDFA (see section 8.1) but with a much smaller geometry (approximately 1 cm or less as opposed to tens of metres for doped fibre amplifiers). Such devices could be used as stand-alone optical amplifiers or, if integrated with silicon photonic components on the same chip, could result in lossless device operation (i.e. the on-chip amplification balancing fibre to chip coupling and waveguide propagation losses). In addition to potential and demonstrated (in the case of Si nano-crystals) gain mechanisms outlined in this chapter, a promising approach is the use of the Raman effect, a potentially straightforward solution which requires no additional device processing above that necessary to fabricate the silicon waveguide.

8.4.1 Spontaneous Raman Effect

The spontaneous Raman effect is a very well known and documented phenomenon covered in a number of textbooks (for example see [25]). It is manifested in the observation of faint sidebands in the frequency spectrum when (almost) monochromatic light scatters from a solid material. The frequency difference between the sidebands and the input power is always fixed and dependent on the make-up of the scattering medium. This can be explained via the transient absorption of the incident photons. In the case of an optical source in the near-infrared (as is generally the case in optical communications) with a wavelength of $1-10\,\mu m$, photon absorption can occur via the excitation of molecular or atomic vibrations of the surrounding medium. This is shown diagrammatically in Figure 8.21.

The initial vibrational state of the molecules or atoms making up the medium is F_p. Upon absorption of an incident photon the system energy is raised to a temporary level E_t until it relaxes to a state E_1, releasing a photon of energy $E_t - E_1$. The excess energy $(E_p - E_1)$ is given up to the system as heat. The transition from E_t to E_1 is known

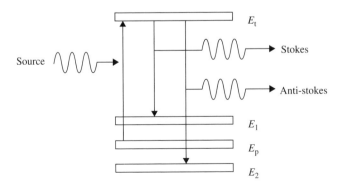

Figure 8.21   Diagram of the spontaneous Raman effect

as a *Stokes transition*. A transition to a level less than E_p (in the above case, E_2), with the subsequent emission of a photon with energy $E_t - E_2$, is known as an *anti-Stokes transition*. Therefore, when the frequency spectrum of the scattered light is measured, in addition to the strong signal corresponding to the frequency of the incident light, the Stokes and anti-Stokes transitions result in fixed sidebands synonymous with the Raman effect.

8.4.2 Stimulated Raman Effect

It is possible to create stimulated Raman scattering (with emission of a coherent light beam) by irradiating a solid with two beams simultaneously, one which excites (or pumps) the constituent molecules or atoms and a second with a wavelength resonant with the Stokes transition. This is shown diagrammatically in Figure 8.22.

In this case, the Stokes transition is stimulated by the signal beam and hence amplification of the signal at an energy of $E_t - E_1$ is observed. Unlike conventional semiconductor laser operation, the amplified wavelength is determined by a combination of the pump wavelength and the energy difference ($E_1 - E_p$), in turn dependent on the scattering medium. The stimulated Raman effect therefore opens the possibility for amplification at a wide range of wavelengths extending from the ultraviolet to the infrared.

8.4.3 Raman Emission from Silicon Waveguides at 1.54 μm

The flexibility of wavelength selection for Raman emission makes it an attractive proposition for integration with silicon waveguide technology.

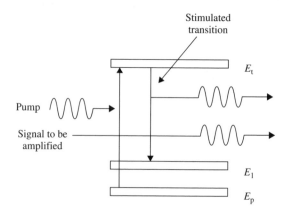

Figure 8.22 Diagram of the stimulated Raman effect

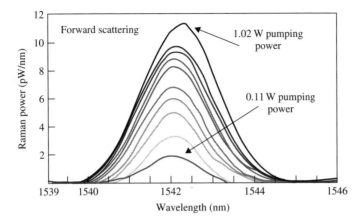

Figure 8.23 Spontaneous Raman spectra measured for different 1.43 μm pumping powers. The greatest intensity shown is for a pumping power of 1.02 W; the least is for 0.11 W. Reproduced from R Claps, D Dimitropoulos, Y Han and B Jalali (2002) 'Observation of Raman emission in silicon waveguides at 1.54 μm', *Optics Express*, **10**, 1305–1313 by permission of Optical Society of America

Recent experimental observation of spontaneous Raman emission from silicon waveguides was reported by the group lead by Jalali at UCLA [26]. A 1.43 μm pumping source was coupled into a silicon-on-SiO₂ rib waveguide structure with a cross-section of approximately 20 μm². The Raman spectra were measured for emission from both end facets of the waveguide. The results for the forward emission (i.e. for light emitted from the opposite end of the waveguide from which the pump signal was introduced) is shown in Figure 8.23. The 1.43 μm pump results in an emission centered at 1.542 μm which varies linearly with the pumping power. The spectral width of the frequency of the emitted light is 105 GHz. The Raman scattering efficiency for silicon was deduced as $4.1 \times 10^{-7}\,\mathrm{cm}^{-1}\cdot\mathrm{Sr}^{-1}$.

In a subsequent publication the same group examined the feasibility of forming a waveguide amplifier using Raman scattering [27]. Using a stimulated gain coefficient of 0.07 cm/MW and accounting for coupling and propagation losses and pump and signal mode mismatch, they determined that a pumping power of around 500 mW at 1.432 μm is required to provide a signal gain of 10 dB for a wavelength of 1.54 μm in a waveguide of 2 cm in length.

A disadvantage of this method for the fabrication of amplifiers is the high power density introduced into the waveguide geometry, which induces nonlinear absorption effects such as two-photon absorption and amplified spontaneous emission.

8.5 SUMMARY

In the first seven chapters we provided an introduction to the field of silicon photonic circuits, but this chapter has offered a glimpse of the future, in one of the many exciting research fields currently associated with silicon photonics. The aim has been to describe work, currently in progress, towards the development of an efficient, silicon-based, light-emitting device operating at wavelengths suitable for use in communication networks. The various approaches described here do not represent an exhaustive list. Most notably, a number of devices incorporating SiGe have been excluded and the reader is referred elsewhere for a description of these approaches [1]. Rather, we have highlighted those technologies which hold greatest promise from the viewpoint of desired functionality and device integration while using techniques for fabrication which are commonplace in a modern silicon fabrication plant.

An efficient silicon light-emitting device (and more significantly a silicon-based laser) is the missing piece in the silicon photonics puzzle. Its development would revolutionise the optoelectronics and micro-electronics industries. Although significant strides have been taken to achieve that dream we are still many years from its realisation. Even the most efficient devices demonstrated to date require unreasonable amounts of power, while device reliability remains an issue. Significant improvements in device performance are required before widespread use becomes feasible. However, the optimist would point to the immense financial rewards that would accompany such a technological development. In our opinion this has already created a critical mass that will eventually result in the development of optical sources that will in turn promote silicon as the leading semiconductor material in the photonics industry.

REFERENCES

1. L Pavesi, S Gaponenko and L Dal Negro (2003) *Towards the First Silicon Laser*, NATO science series, Kluwer Academic Publications, Dordrecht.
2. P J Mears, L Reekie, I M Jauncy and D N Payne (1987) 'Low-noise erbium-doped fibre amplifier operating at 1.54 µm', *Electron. Lett.*, **23**, 1026–1027.
3. A Polman (1987) 'Erbium-implanted thin film photonic materials', *J. Appl. Phys.*, **82**, 1–39.
4. Y H Xie, E A Fitzgerald and Y J Mii (1991) 'Evaluation of erbium-doped silicon for optoelectronic application', *J. Appl. Phys.*, **70**, 3223–3228.

5. A Polman, J S Custer, E Snoeks and G N van den Hoven (1993) 'Incorporation of high concentrations of erbium in crytal silicon', *Appl. Phys. Lett.*, **62**, 507–509.

6. A Polman, G N van den Hoven, J S Cluster, J H Shin and R Serna (1995) 'Erbium in crystal silicon: optical activation, excitation, and concentration limits', *J. Appl. Phys.*, 77, 1256–1262.

7. S Coffa, G Franzò, F Priolo, A Polman and R Serna (1994) 'Temperature dependence and quenching processes of the intra-4f luminescence of Er in crystalline Si', *Phys. Rev. B.*, **49**, 16313–16320.

8. S M Sze (1981) *Physics of Semiconductor Devices*, John Wiley & Sons, New York.

9. S Libertino, S Coffa and M Saggio (2000) 'Design and Fabrication of integrated Si-based optoelectronic devices', *Mat. Sci. Semicond. Process.*, 3, 375–381.

10. B Zheng, J Michel, F Y G Ren et al. (1994) 'Room-temperature sharp line electroluminescence at $\lambda = 1.54\,\mu$m from an erbium-doped, silicon light emitting diode', *Appl. Phys. Lett.*, **64**, 2842–2844.

11. G Franzò, S Coffa, F Priolo and C Spinella (1997) 'Mechanism and performance of forward and reverse bias electroluminescence at $1.54\,\mu$m from Er-doped Si diodes', *J. Appl. Phys.*, **81**, 2784–2793.

12. L T Canham (1990) 'Silicon quantum wire array fabrication by electrochemical and chemical dissolution of wafers', *Appl. Phys. Lett.*, 57, 1046–1048.

13. A Richter, P Steiner, F Kozlowski and W Lang (1991) 'Current-induced light emission from a porous silicon device', *IEEE Electron. Device Lett.*, **12**, 691–692.

14. B Gelloz and N Koshida (2000) 'Electroluminescence with high and stable quantum efficiency and low threshold voltage from anodically oxidised thin porous silicon diode', *J. Appl. Phys.*, 88, 4319–4324.

15. A G Cullis, L T Canham and P D J Calcott (1997) 'The structural and luminescence properties of porous silicon', *J. Appl. Phys.*, **82**, 909–965.

16. F Iacona, G Franzò and C Spinella (2000) 'Correlation between luminescence and structural properties of Si nanocrystals', *J. Appl. Phys.*, 87, 1295–1303.

17. K S Min, K V Shcheglov, C M Yang et al. (1996) 'Defect-related versus excitonic visible light emission from ion beam synthesized Si nanocrystals in SiO_2', *Appl. Phys. Lett.*, **69**, 2033–2035.

18. A Irrera, D Pacifici, M Miritello et al. (2003) 'Electroluminescence properties of light emitting devices based on silicon nanocrystals', *Physica E*, **16**, 395–399.

19. L Pavesi, L Dal Negro, G Mazzoleni, G Franzò and F Priolo (2000) 'Optical gain in silicon nano-crystals', *Nature*, **408**, 440–444.

20. L Pavesi (2003) 'A review of the various efforts to a silicon laser', in D J Robbins and G E Jabbour (eds), *Optoelectronic Integration on Silicon*, Proceedings of SPIE (Photonics West).

21. G Franzò, D Pacifica, V Vinciguerra, F Priolo and F Iacona (2000) 'Er3+ ions-Si nanocrystals interactions and their effects on the luminescence properties', *Appl. Phys. Lett.*, **76**, 2167–2169.
22. M E Castagna, S Coffa, M Monaco et al. (2003) 'Si-based materials and devices for light emission in silicon', *Physica E*, **16**, 547–553.
23. W L Ng, M A Lorenço, R M Gwilliam et al. (2001) 'An efficient room-temperature silicon-based light emitting diode', *Nature*, **410**, 192–194.
24. M A Lorenço, M S A Siddiqui, R M Gwilliam, G Shao and K P Homewood (2003) 'Efficient silicon light emitting diodes made by dislocation engineering', *Physica E*, **16**, 376–381.
25. E Hecht (2002) *Optics*, Addison-Wesley, San Francisco.
26. R Claps, D Dimitropoulos, Y Han and B Jalali (2002) 'Observation of Raman emission in silicon waveguides at 1.54 μm', *Optics Express*, **10**, 1305–1313.
27. R Claps, D Dimitropoulos and B Jalali (2002) 'Stimulated Raman scattering in silicon waveguides', *Electron. Lett.*, **38**, 1352–1354.

Index

Silicon Photonics: An Introduction   Graham T. Reed and Andrew P. Knights
© 2004 John Wiley & Sons, Ltd   ISBN: 0-470-87034-6